D1631931

ISLANDS OF SCOTLAND

Islands of the British Isles

ISLANDS OF SCOTLAND

ISLANDS OF IRELAND

ISLANDS OF ENGLAND AND WALES

ISLANDS OF SCOTLAND

Donald McCormick

Osprey

First published in 1974 by
Osprey Publishing Ltd.,
707 Oxford Road, Reading, Berkshire

ISBN 0 85045 167 1

Filmset and printed in Monophoto Plantin by
BAS Printers Limited, Wallop, Hampshire

INTRODUCTION

A distinctive feature of the islands of the British Isles and Ireland is that most of them are situated off the west coasts. In Scotland this feature is even more pronounced, there being several hundred isles off the western shores, but not even a score off the eastern seaboard.

Climatically this is fortunate, as the weather is much milder in the west, and even in the Hebrides one can find a wealth and variety of flowers such as could not possibly grow on the bleaker, more exposed islands of Orkney and Shetland to the far north.

It would probably take several years to explore thoroughly the islands of Scotland. The Outer Hebrides alone extend for more than a hundred miles from the Butt of Lewis in the north to Vatersay in the south, and in Lewis and Harris they possess the largest island in the British Isles. There is, perhaps, greater variety in the Inner Hebrides, ranging from the mountains of Rhum to the evergreen valleys and fields of Skye with its intermittent swirling mists and dazzling rainbows and sunsets.

So numerous are the islands of Scotland that many of them – even those with a fair acreage – are relatively unknown and largely unexplored. This is particularly true of a number of islands in both the Orkneys and the Shetlands, and in the north-west of Scotland. It will be noted that some of these are listed individually but only briefly, giving their situation, size, etc., but providing no further details.

Had these been tiny islands with nothing particularly significant to report on them, their names would have been included in the entry on the group of islands to which they belong. However, they have been listed individually because, although there is little to say about such uninhabited and unmemorable isles, their size alone may tempt someone to explore further. No doubt some of them retain undiscovered archaeological remains.

When in doubt about including such terse entries I have thought of that phrase of Robert Browning's, 'some unsuspected isle in far-off seas'. My quest for islands often tempted me to include such 'unsuspected' isles, sometimes mere rocks or sandbanks, which for some reason or other captured my imagination.

Into this category come both Rockall, an obvious rock of modest dimensions and yet of great importance, and Kisimul, surely the tiniest isle in the book and to all intents and purposes simply a castle rising out of Castlebay, Barra.

For the purpose of this book I have omitted all islands in inland lochs and those in the upper reaches of rivers. I have, however, included Mugdrum in the Firth of Tay and islands in the Firth of Clyde and the long arm of Loch Linnhe where it runs down to the sea.

I should be the last to deny that there may well be some 'unsuspected' isles which I have ignored, but which deserve inclusion. But I should like to stress that some of these, while not listed separately, are included under the groups of islands to which they belong. Thus the reader should take care to note that there are the Barra Isles as well as Barra, and that the delightfully-named Summer Isles include several attractive islets not all of which are individually listed.

Some quite large islands have been given relatively little space, while some small ones seemed to deserve more space than their size might warrant. Staffa is perhaps one of the best examples of tiny islands which provide not only much to write about, but much to inspire one, as Mendelssohn found when he composed his 'Hebrides' overture there.

I have tried to give information for a wide range of

island-lovers: how to get there; the problems of finding a landing-place (this is often as difficult in the Inner Hebrides as in the Outer); warnings of what to avoid (Gruinard is banned to visitors); details of accommodation where this is available; climate (avoiding the traditionally rainy months is important on west-coast islands); and recreational facilities (even in Orkney and Shetland there are some excellent golf-courses). I have also tried to indicate the scope for people with special interests: bird-watchers (there is a Bird Observatory on Fair Isle); archaeologists (for whom Orkney is a treasure house); botanists (Gigha's lovely gardens are outstanding); historians; climbers (for the very bold the Old Man of Hoy is the supreme test); fishermen; yachtsmen; geologists (Arran is especially noteworthy); people in quest of solitude; those whose tastes are for *ceilidh* concerts in remote Gaelic outposts (Barra); collectors of unusual stamps (Staffa); pot-holers; under-water swimmers; marine biologists (the Cumbraes); animal-lovers; the ordinary tourist who wants something different; hikers; honeymooners and romantics of every kind.

On the subject of romance, a curious feature of many Hebridean islands is the quantity of legends associating them with the fairies. I found Luchruban (Pigmies' Isle) fascinating if only because of the stories of the 'little people' who had inhabited it. Many of the islands, as around the English, Welsh, and Irish coasts, are named after saints, bearing witness to the fact that in the Dark Ages it was the tiny, remote islands off our coasts which alone bravely and stoically maintained traditions of Christianity and scholarship.

The islands around Scotland are richly steeped in history, legend, and literature. If some of the legends are far-fetched, they are nevertheless often substantiated by some iota of factual evidence – even the small (human?) bones found on Pigmies' Isle. The literature relating to the Scottish islands is very diverse, from *Macbeth*, in which Shakespeare refers to Inchcolm in the Firth of Forth, to Sir Compton Mackenzie's *Whisky Galore*, a book based originally on a shipwreck off the tiny isle of Calvay, and from S. R. Crockett's stories of 'Grey Galloway' to the poets of Orkney and Shetland, and that romantic lover of islands, Robert Louis Stevenson. If so minded, you can retrace the journey of David Balfour to the Isle of Erraid where he was marooned in *Kidnapped*.

One must constantly revise a book such as this, not so much because erosion and changes in tidal currents cause islands to disappear (this is less the case with Scottish islands than it is around the English coast), but because population changes can be rapid. The drift from the remoter islands continues remorselessly though, as will be seen, there are always a few adventurers each year who are prepared to start life afresh on some uninhabited isle. It has not always been easy to get accurate population figures because some islands listed as uninhabited have seasonal residents, while an island which has eighty inhabitants one month might be reduced to a mere score by a mass exodus a few months later. The figures given are for 1961 unless otherwise stated.

In concluding I should like to acknowledge the help I received in compiling this work, and to express my gratitude to the Royal Geographical Society; the National Trust for Scotland; the Scottish Tourist Office; the Scottish National Library in Edinburgh; Miss Diane Fisher, for research on charts and maps; my son, A. S. McCormick, for various reports; and some of the private owners of islands, too numerous to mention individually.

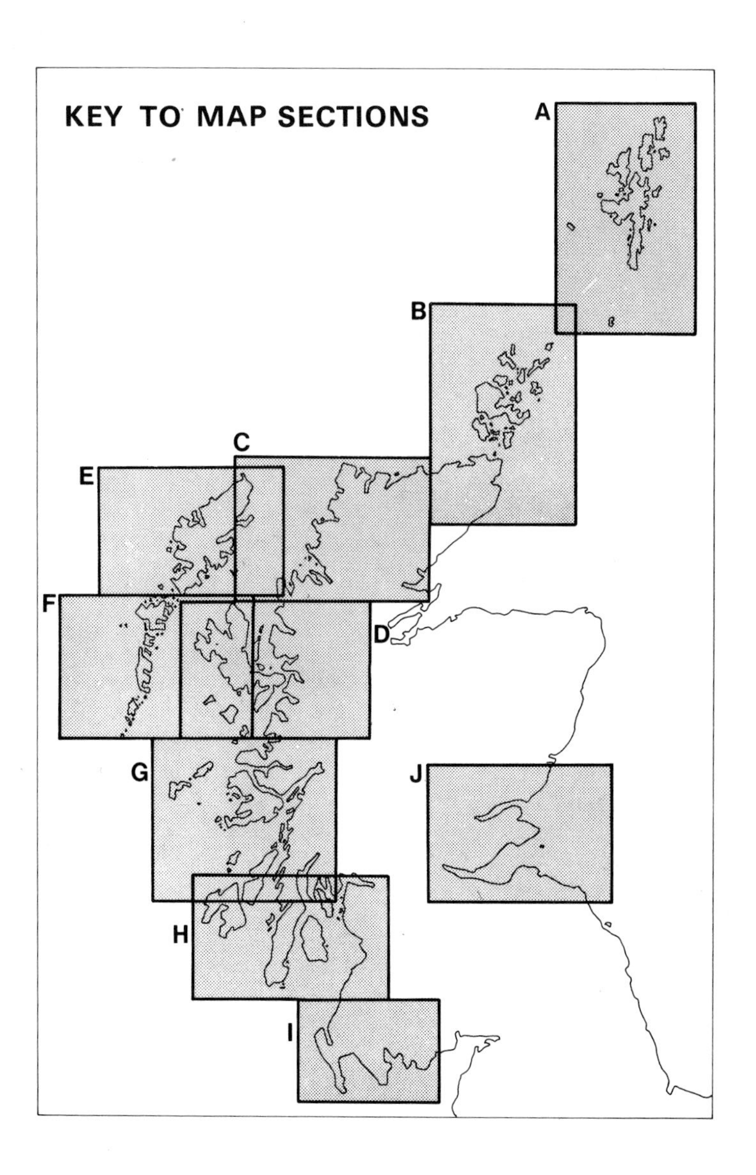
KEY TO MAP SECTIONS
A
B
C
D
E
F
G
H
I
J

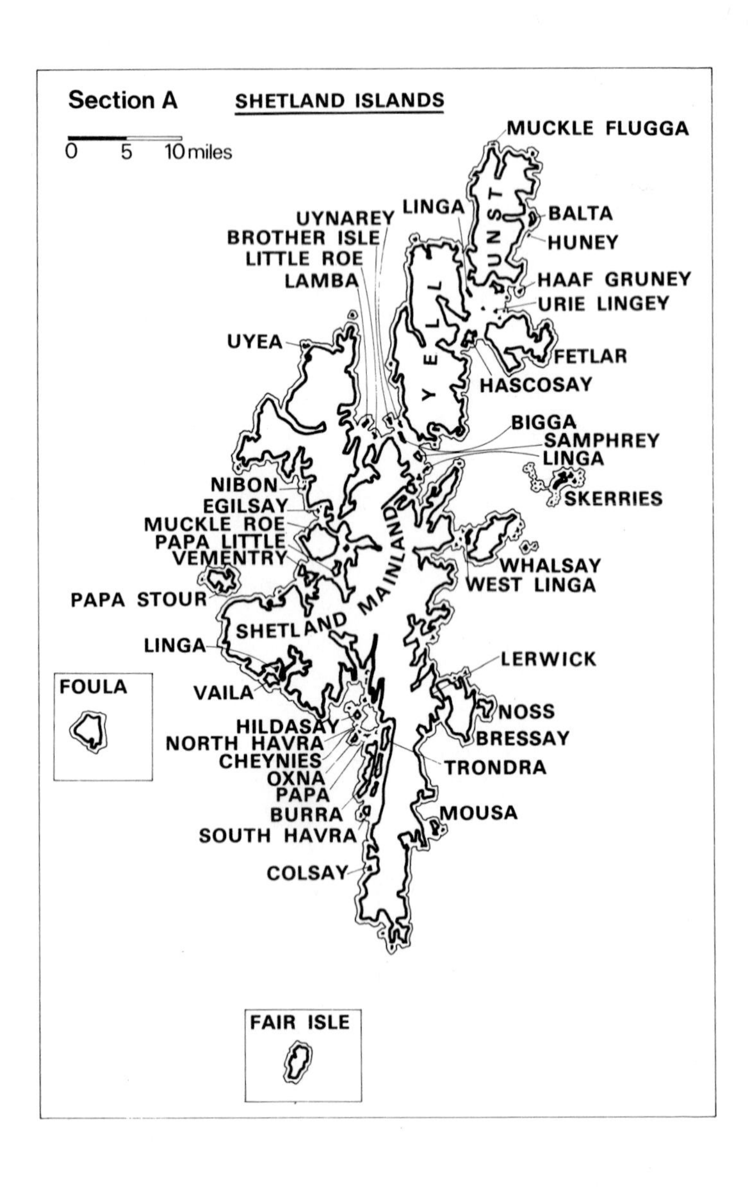
Section A
SHETLAND ISLANDS
0 5 10 miles
MUCKLE FLUGGA
UNST
BALTA
HUNEY
HAAF GRUNEY
URIE LINGEY
FETLAR
LINGA
UYNAREY
BROTHER ISLE
LITTLE ROE
LAMBA
YELL
HASCOSAY
UYEA
BIGGA
SAMPHREY
LINGA
SKERRIES
NIBON
EGILSAY
MUCKLE ROE
PAPA LITTLE
VEMENTRY
WHALSAY
WEST LINGA
PAPA STOUR
SHETLAND MAINLAND
LINGA
LERWICK
FOULA
VAILA
NOSS
HILDASAY
NORTH HAVRA
BRESSAY
CHEYNIES
TRONDRA
OXNA
PAPA
BURRA
MOUSA
SOUTH HAVRA
COLSAY
FAIR ISLE

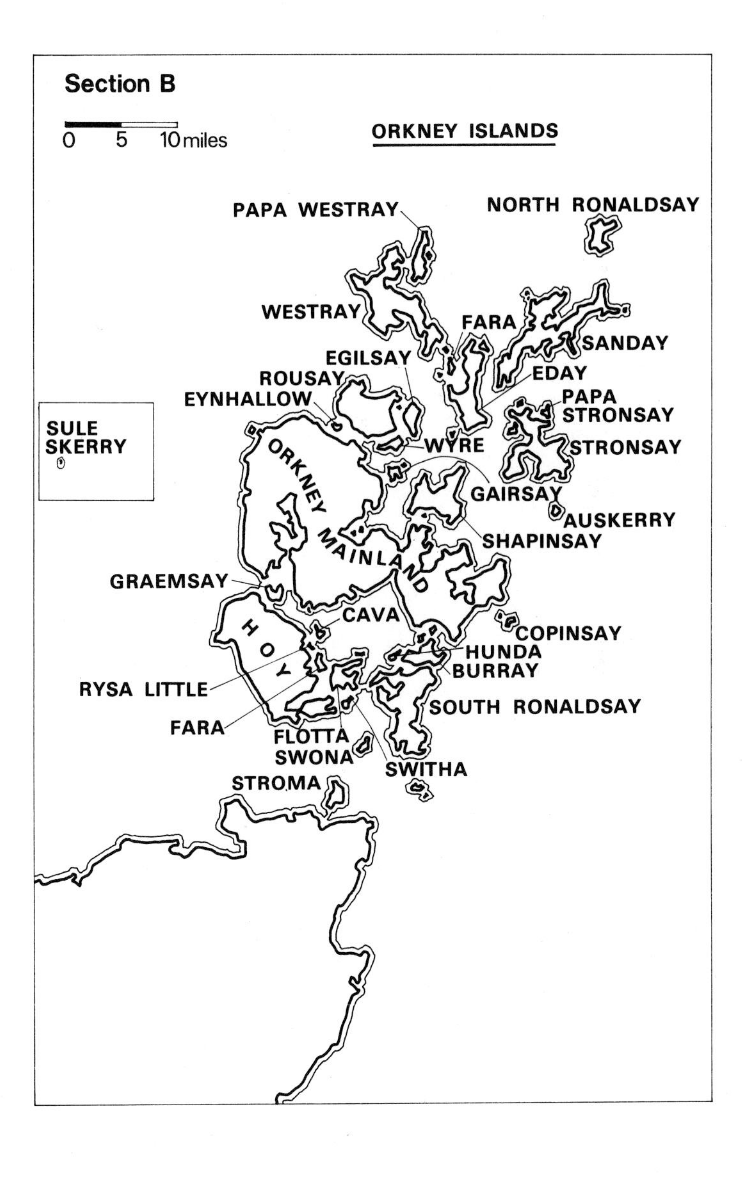
Section B
0 5 10 miles
ORKNEY ISLANDS
PAPA WESTRAY
NORTH RONALDSAY
WESTRAY
FARA
SANDAY
EGILSAY
EDAY
ROUSAY
PAPA STRONSAY
EYNHALLOW
SULE SKERRY
WYRE
STRONSAY
ORKNEY MAINLAND
GAIRSAY
AUSKERRY
SHAPINSAY
GRAEMSAY
CAVA
HOY
COPINSAY
HUNDA
BURRAY
RYSA LITTLE
SOUTH RONALDSAY
FARA
FLOTTA
SWONA
SWITHA
STROMA

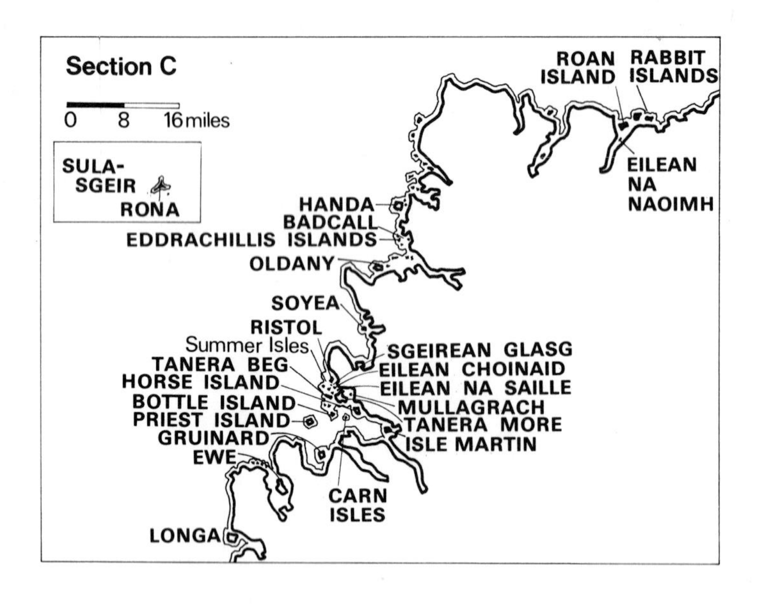
Section C
0 8 16 miles
SULA-
SGEIR
RONA
ROAN
ISLAND
RABBIT
ISLANDS
EILEAN
NA
NAOIMH
HANDA
BADCALL
EDDRACHILLIS ISLANDS
OLDANY
SOYEA
RISTOL
Summer Isles
TANERA BEG
HORSE ISLAND
BOTTLE ISLAND
PRIEST ISLAND
GRUINARD
EWE
SGEIREAN GLASG
EILEAN CHOINAID
EILEAN NA SAILLE
MULLAGRACH
TANERA MORE
ISLE MARTIN
CARN
ISLES
LONGA

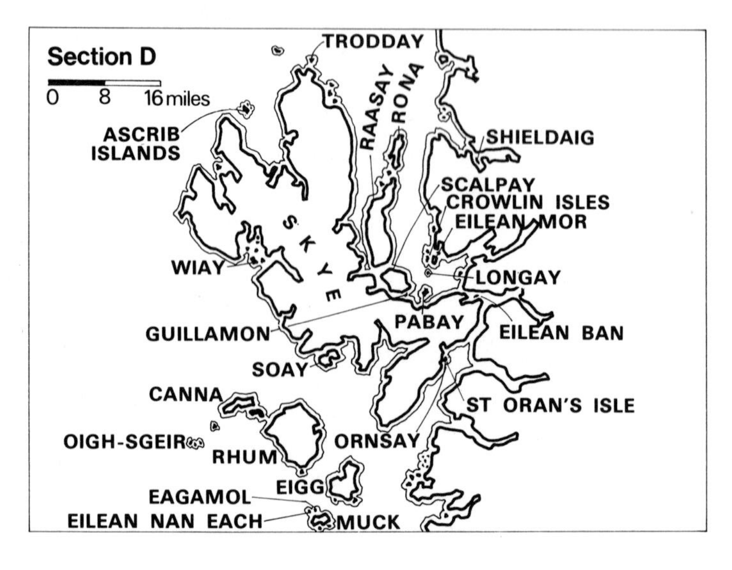
Section D
0 8 16 miles
TRODDAY
ASCRIB
ISLANDS
RAASAY
RONA
SHIELDAIG
SCALPAY
CROWLIN ISLES
EILEAN MOR
S K Y E
WIAY
LONGAY
PABAY
EILEAN BAN
GUILLAMON
SOAY
CANNA
ST ORAN'S ISLE
OIGH-SGEIR
ORNSAY
RHUM
EIGG
EAGAMOL
EILEAN NAN EACH
MUCK

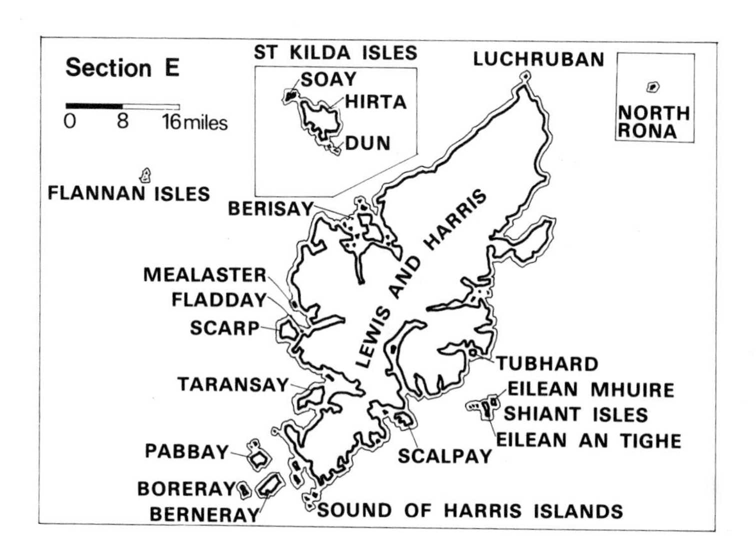
Section E
0
8
16miles
ST KILDA ISLES
SOAY
HIRTA
DUN
LUCHRUBAN
NORTH
RONA
FLANNAN ISLES
BERISAY
LEWIS AND HARRIS
MEALASTER
FLADDAY
SCARP
TUBHARD
TARANSAY
EILEAN MHUIRE
SHIANT ISLES
EILEAN AN TIGHE
PABBAY
SCALPAY
BORERAY
BERNERAY
SOUND OF HARRIS ISLANDS

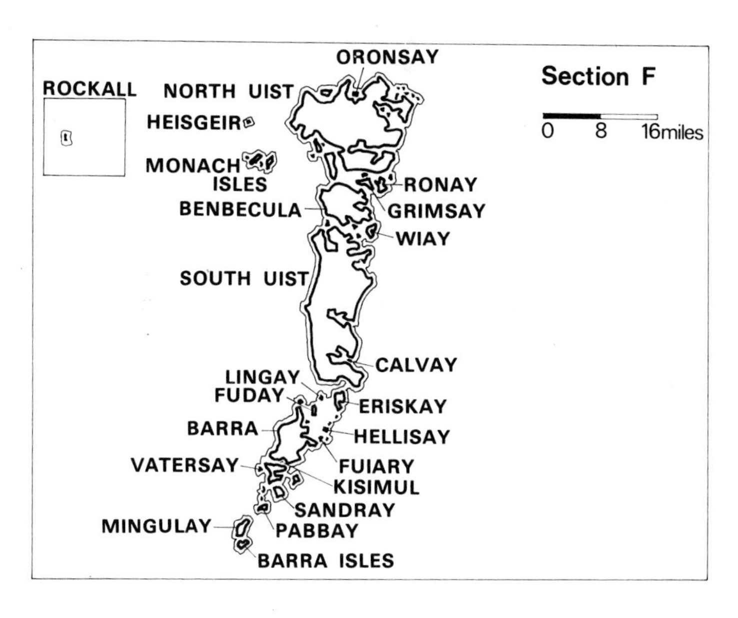
ORONSAY
Section F
0
8
16miles
ROCKALL
NORTH UIST
HEISGEIR
MONACH
ISLES
RONAY
BENBECULA
GRIMSAY
WIAY
SOUTH UIST
CALVAY
LINGAY
FUDAY
ERISKAY
BARRA
HELLISAY
VATERSAY
FUIARY
KISIMUL
SANDRAY
MINGULAY
PABBAY
BARRA ISLES

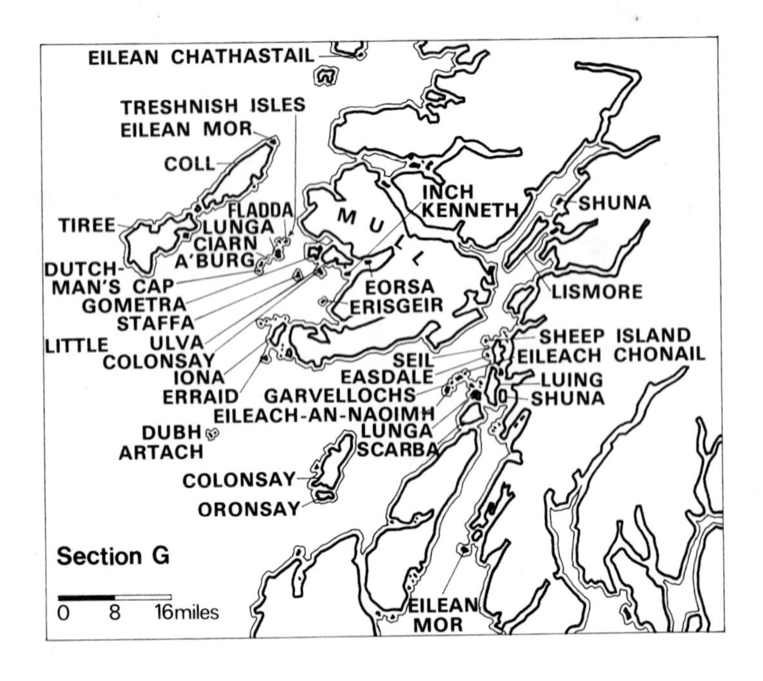

Section G

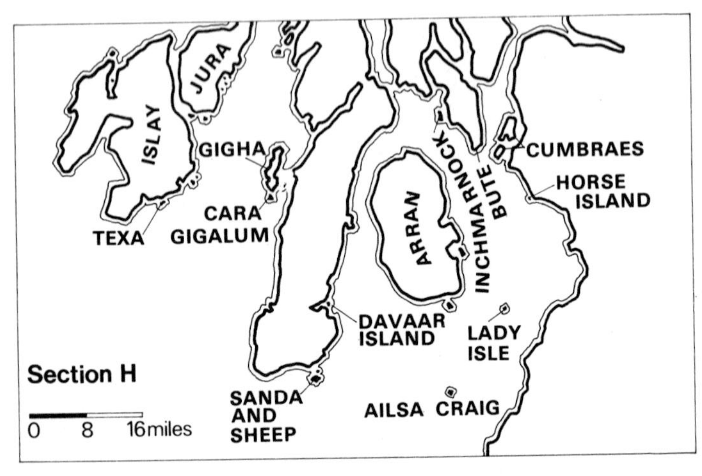

Section H

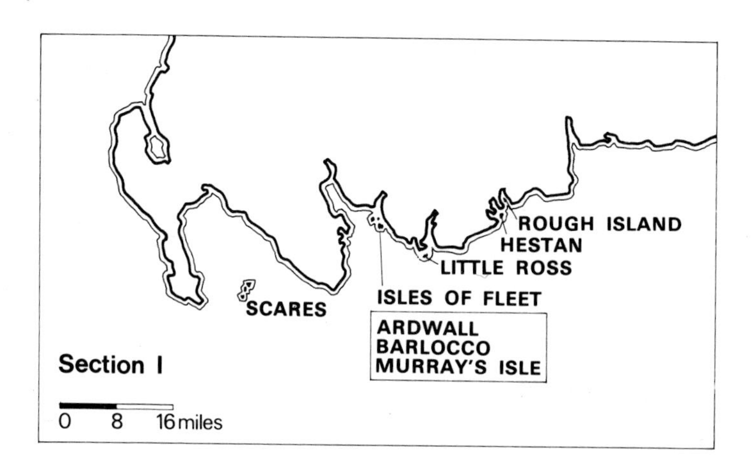
ROUGH ISLAND
HESTAN
LITTLE ROSS
ISLES OF FLEET
SCARES
ARDWALL
BARLOCCO
MURRAY'S ISLE
Section I
0 8 16 miles

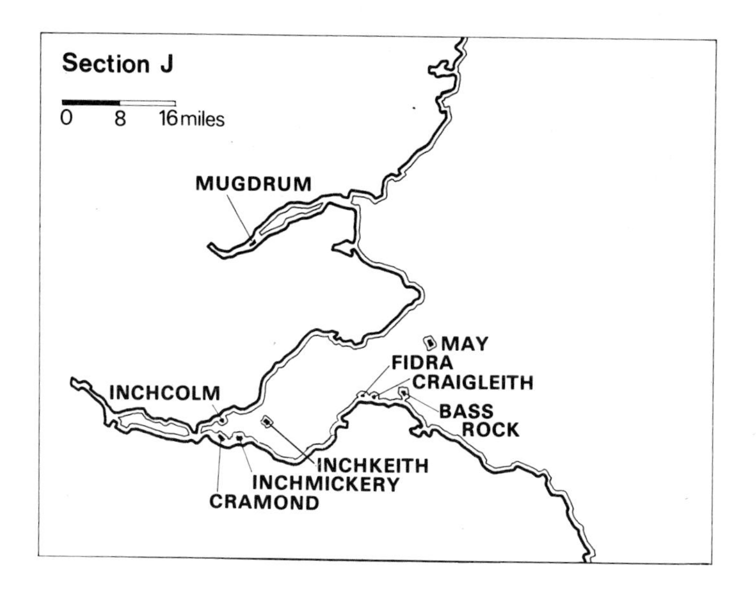
Section J
0 8 16 miles
MUGDRUM
MAY
FIDRA
CRAIGLEITH
INCHCOLM
BASS
ROCK
INCHKEITH
INCHMICKERY
CRAMOND

AILSA CRAIG

SITUATION: 10 miles W. of Girvan, Ayrshire.
AREA: circumference of 2 miles.
POPULATION: uninhabited.
ACCESS: special arrangements can be made to visit the island from Girvan by boat.

Often referred to as 'Paddy's Milestone' because it lies on the sea route between Scotland and Ireland, Ailsa Craig is a prominent sea-mark for many miles around. Its gaunt granite peak rises to a height of 1,114 ft, and its gannetry and colonies of puffins, guillemots, and other sea birds make it a special attraction to ornithologists. Otherwise it is a bleak, forbidding place. Its granite is used to make curling stones.

AN CEANN EAR: *see under* Monach Isles.

AN CEANN IAR: *see under* Monach Isles.

ARDWALL: *see under* Isles of Fleet.

ARRAN

SITUATION: in the Firth of Clyde 20 miles W. of Ayr and 4 miles E. of the Mull of Kintyre.
AREA: 166 sq. miles.
POPULATION: 4,094 (1970).
ACCESS: by steamer from Ardrossan; car ferry service on Sundays in summer only.

opposite: *Arran: Glen Rosa*

Arran (in the county of Bute) is so varied in its scenery that it can be said to be almost a Scotland in miniature. There are mountains in the N., moors in the interior, sheltered valleys, lochs, splendid beaches, and gardens filled with flowers. It is still largely unspoilt and is encircled by a road that takes in some magnificent views.

It dominates the Firth of Clyde, and towering above all else from a distance is the peak of Goat Fell (2,866 ft), surrounded by the only slightly smaller mountains of Caisteal Abhail (2,817 ft) and Beinn Tarsunn (2,706 ft). Goat Fell is a fairly easy climb if taken from the direction of Brodick, Arran's chief town. On a clear day England, Ireland, and the Isle of Man can be seen from the top of this mountain.

The road round the island only leaves the coast briefly at points in the N. and S. Two other roads cross the island

Arran: Lochray from the south

from E. to W., and it is at various points on these that one can explore the wild, untamed, and largely unsettled interior of Arran. Most of the villages are along the coast; there are fourteen of them altogether, if one includes Lamlash, which is almost a town. Most of the small farms lie close to the coast. The three principal bays of the island – Brodick, Lamlash, and Whiting – all cater for holiday-makers.

James Hutton, the geologist who visited Arran in the 18th century, declared that the granite mountains of Arran were the eroded remnants of a great mass of molten material which was poured out of the earth. Characteristic features of the landscape are the basaltic and other dikes which exist everywhere. Outstanding among these is the Witch's Step, a huge V-shaped chasm which breaks the ridge between Caisteal Abhail and Suidhe Fhearghas. There is a great deal of literature on the geology of Arran, and it is still visited by geologists from all over the world, not merely to study, but to take samples.

The lochs are mainly situated on high land, the largest being Loch Tanna, a mile long and half a mile wide, which is drift-dammed and shallow. The most impressive of the rock basins are Loch Chorein Lochain in the NW., surrounded by steep granite rocks, and Loch Garbad in the S.

Arran's climate is mild, thanks to the Gulf Stream, but owing to the prevailing SW. wind blowing in from the Atlantic, its rainfall is high, the E. coast being wetter than the W. Vegetation is varied: there are beech, chestnut, elm, lime, larch, and Scottish pine trees, and on the high ground masses of heather, lichens, mosses, and alpine willow, many species of berry, such rare plants as purple and starry saxifrage and, lower down the slopes, ferns of every conceivable type.

Most Arran cottages have a few fruit trees in their gardens, plum or apple, with some currant bushes and occasionally a bed of strawberries or a row of raspberry canes. Flowers are to be seen at their best in Brodick

Castle Gardens. There are two gardens here: a woodland garden started in 1923 by the late Duchess of Montrose, which has a splendid display of rhododendrons and hydrangeas, and a formal garden dating from 1710. Here are magnolias, camellias, and palm lilies, a type imported from New Zealand. Brodick Castle was the ancient seat of the Dukes of Hamilton, and more recently of Mary, Duchess of Montrose. In Viking times a fortress was sited here and part of the present castle dates from the 14th century. It was extended in 1652 and 1844. Its contents include silver, porcelain, and paintings from the collections of the Dukes of Hamilton, William Beckford, and the Earls of Rochford. In 1958 the castle and grounds were taken over by the National Trust for Scotland, and Lady Jean Fforde, daughter of Mary, Duchess of Montrose, made a gift to the Trust of 7,300 acres of mountainous country on the island, including Goat Fell.

The most attractive of the wild animals of Arran are the red deer which have been on the island for centuries: Caolite, a Celtic poet, referred to 'Arran of the many stags'. Towards the end of the 18th century indiscriminate slaughter threatened the red deer with extinction, but the total number today is said to be about 1,600. There are red squirrels on the island, especially in the woods at Brodick, but so far no grey squirrels. Otters are to be seen quite frequently, and seals can also be found around the rocky coast of Arran. Of domestic animals the principal are sheep, but there are several herds of crossbred cattle, and poultry are abundant.

Occasionally the golden eagle can be seen high up in the mountains, but more common birds of prey are the buzzard and sparrowhawk. A variety of gulls can be found near the lochs, and ravens nest in the N. of the island, while in the S. there are swallows, house-martins, sand-martins, blackbirds, long-tailed tits, water ousel, kingfisher, and sand-piper. Grouse and woodcock are the chief game birds, but their numbers are declining.

Fishing is better in the SW. than elsewhere. Here there are salmon- and trout-fishing and the salmon are said to average 9 lb. Around the coast cod, haddock, and whiting are plentiful, and in July shoals of mackerel appear.

There is evidence that there were coastal Mesolithic people in Arran, and the island was inhabited during the Neolithic age. A survey of the chambered cairns of Arran was made by Thomas Bryce in 1910 – he examined seventeen examples, the best being also the most remote, the Carn Ban, 950 ft above sea level. This cairn was 100 ft long and 60 ft wide, and contained the remains of a semi-circular forecourt out of which led a portalled chamber which was divided into four compartments. Human remains have been found in some of the cairns. In the King's Cave in the red sandstone cliffs near Drumadoon there are faint carvings, including one of a cross rising out of foliage, which suggests that the cave may have been used as a chapel or a hermit monk's cell at one period.

The Arran forts are in various parts of the island, and they are of differing shapes and sizes – circular, oval, and oblong. Archaeologists have established that some of them were built about 200 BC.

St Brendan, that seafaring missionary, is said to have visited Arran and given his name to Kilbrannan Sound, but there is no proof of this. It is also uncertain at just what period Arran was first visited by the early Christian missionaries, though a monastery was set up at Kilpatrick about the end of the 6th century. There are several Norse place-names in Arran, both past and present, which testify to an occupation of the island by the Scandinavian hordes. A notable example is Geita-fjall (Goat Fell).

The Norse invaders mated with the native Gaels, and in due course their offspring began to challenge the rulers of Norway. Galleys from the Isle of Arran certainly took part in the sea battle in 1156 when the Scots defeated the Norwegians. Following this Arran was rent for many years by a series of clan feuds, mainly between the Campbells and the Macdonalds. But eventually it was the Stewarts who became the Lords of Arran. In 1357, it is recorded, the Stewart lord of Knapdale and Arran granted 'the churches of St Mary and St Brigid in the island of Arran' to the monastery of Kilwinning in Ayrshire.

After the death of Alexander III of Scotland, Edward I of England also claimed overlordship of Scotland, and granted Arran to one Thomas Bysset. Robert Bruce, who had had to flee from the mainland because he had failed to find sufficient support to mount a war against the English, went to Arran in the hope of rallying enemies of Bysset. Bruce landed in Arran with thirty small galleys – in the words of the chronicler of his day:

> The King arrivit in Arane;
> And syne to the land is gane,
> And in a toune tuk his herbery.

He probably landed in Whiting Bay, but it is less certain that he hid in the so-called King's Cave at Drumadoon, as legend has it. Certainly Robert Bruce remained on Arran for several days, if not weeks, and probably won a certain amount of support there.

After the Battle of Bannockburn in 1314, when the issue of the independence of Scotland was settled by a rousing Scottish victory, Arran once again became the property of the Stewarts. In 1609 a commission of justiciary over the whole island was conferred on the first Marquis of Hamilton, and the island remained in the family until the twelfth Duke's death, after which it passed to his only daughter who became Duchess of Montrose.

Some attempts were made to set up industries on Arran in the 19th century. Then there were carding mills at Brodick and Burican, a flax mill at Lagg, and a wauking and dye mill at Monachmore. But these, like the mining of coal and quarrying of slate in the Lochranza area, failed before the end of the century. The red sandstone quarries at Brodick and elsewhere were closed in 1928. Barytes was mined at Sannox as early as 1840 and achieved a production peak of 9,000 tons per year in the 1930s, but just before World War II the vein was finished, and the pier and light railway serving it were eventually dismantled.

Tourism came to Arran towards the end of the 18th century, and to some extent it was sustained in the following century by Sir Walter Scott's using the island as a background for his poem *The Lord of the Isles*, published in 1815. Robert Browning visited the island in 1862, and Lewis Carroll came here in quest of an artist for his books. Today the holiday trade is considerable. All sailings to the island are linked by buses to the various villages, and there is ample accommodation. In some of the villages there are typical 'tourist shops' catering for visitors with local knitwear, pottery, and hand-wrought jewellery made from such local stones as cornelian, smoky quartz, agate, and amethyst. A wide range of amenities is available for visitors, from speed-boats and water-skiing at Blackwater and Brodick, pony trekking, and sea angling, to tennis, bowls, and golf. There are eighteen-hole golf courses at Brodick, Lamlash, and Whiting Bay, and other smaller courses at Lochranza and Sannox. Highland Games are held at Brodick during the first week of August, and the annual sheep-dog trials at Machrie take place each July.

In 1967 two London University lecturers in geography suggested in a report to the Arran Council of Social Service that the best future for Arran was as an island of retirement. They forecast that the population may be down to no more than 2,000 by 1980, pointing out that it had already fallen to half its peak of nearly 7,000.

See also Holy Island (Arran), Pladda.

Accommodation: there are 67 hotels and boarding-houses on Arran, of which 12 have full licences. Information on these can be obtained from the Isle of Arran Tourist Association, Lamlash. Several hundreds of cottagers also let rooms, and there are usually a number of houses and cottages to let furnished.

Churches: there are Church of Scotland, Free Church, and Congregational churches throughout the island, and Roman Catholic services are held in summer in Brodick public hall.

Places of interest: Brodick Castle, which is 1½ miles from

Brodick Pier, is open from Easter to September, 1.00 pm to 5.00 pm daily; except Sundays, 2.00 pm to 4.30 pm. Admission is to castle and gardens or to gardens only (10 am to 5 pm daily).

Books: *The Isle of Arran*, by Robert McLellan, David & Charles, 1970; *The Geology of Arran*, by G. W. Tyrrell, Edinburgh University Press, 1928; *Arran with Camera and Sketchbook*, by V. A. Firsoff, Robert Hale, 1951; *All About Arran*, by R. A. Downie, Blackie, 1933; 'Fort, Brochs and Wheel-Houses in Northern Scotland', by J. R. C. Hamilton in *The Iron Age in North Britain* (Ed. Rivet), Edinburgh University Press, 1966; *Rock Climbs in Arran*, by J. M. Johnstone, Scottish Mountaineering Club, 1958.

ASCRIB ISLANDS

SITUATION: at the mouth of Loch Snizort, Skye, 2½ miles from the coast of Skye.
AREA: 100 acres.
POPULATION: uninhabited.
ACCESS: by boat from Uig Bay, Skye.

A group of tiny islands occasionally used for grazing, the Ascribs number five, the largest of which are Ascrib itself in the s. of the group, and Iosa.

AUSKERRY

SITUATION: in the Orkney Islands, 4 miles s. of Stronsay.
AREA: 1¼ miles by ½ mile.
POPULATION: uninhabited.
ACCESS: by boat from Lerwick.

BADCALL

SITUATION: in Eddrachillis Bay, Sutherland, close inshore to the Scottish mainland 2 miles from Badcall village at the N. end of the bay.
AREA: 200 acres.
POPULATION: uninhabited.
ACCESS: by boat from the mainland.

Badcall is really a cluster of islands, sometimes used for grazing, belonging to the scattered group of islands in Eddrachillis Bay. Many of the islets are mere rocks, some of which are submerged at high water. They extend about 2 miles NE. from Meall Mor and Meall Beag, the larger isles being in the NE. – Eilean na Rainich, Ceannamhor, Eilean na Bearachd, and Eilean Garbh. They are not as prolific in bird life as Handa (q.v.) to the N., but nevertheless have some interest for the ornithologist.

Eilean na Rainich is 55 ft high and is covered with heather. Ceannamhor is 138 ft and has patches of grass and heather. Eilean na Bearachd also reaches a peak of 138 ft and is covered with heather, while Eilean Garbh consists of two islands almost joined together, and reaching 102 ft and 74 ft respectively.

See also Eddrachillis Islands.

BALTA

SITUATION: 1 mile E. of Unst in the Shetlands, at the entrance to Balta Sound.
AREA: 231 acres.
POPULATION: uninhabited.
ACCESS: by boat from Unst.

Until the latter part of the 19th century Balta was the scene of great activity, as the fishing fleet used to assemble in the Sound close to the island. Many small craft put in at Balta at that time, but it never seems to have been inhabited, even when the herring-curing industry was concentrated up here.

See also Shetland Isles.

BARLOCCO: *see under* Isles of Fleet.

BARRA

SITUATION: principal island of the Barra Isles, lying off the S. tip of the Outer Hebrides, 108 miles from Oban and 8 miles S. of South Uist.

AREA: 20 sq. miles.

POPULATION: 1,750 (1970).

ACCESS: by boat from Oban to Castlebay: Mondays, Wednesdays, and Fridays, leaving Oban at 7.00 am (the voyage takes 10 hrs); or by air from Glasgow daily.

This is an island with such beautiful views that Sir Compton Mackenzie, when he once made a home there, enshrined one of them in a vast, gilded picture-window. It is a lovely island with vivid contrasts in colour – soft, varied, silky greens, sunsets that fade from orange into purple, hills that are mysteriously blue, and sands of shimmering silver.

There is a road round the coast that enables the visitor to enjoy a variety of views, from the cockle strand in the N., where the daily plane from Glasgow lands at low tide, to the trout-filled lochs of the interior. The highest point of the island is Heaval (1,260 ft), which lies to the N. of Castlebay. The latter is the chief town of the island, and

Barra: Breivig Bay (Aerofilms)

has a pier where the steamers land their passengers, a hotel, some shops, and a 19th-century Catholic Church. The inhabitants, who are mainly crofters and fishermen, are nearly all Catholics – there are three Catholic churches and only one belonging to the Church of Scotland on the island.

There is nothing specially distinctive about the architecture of Castlebay, but it makes a good centre for seeing the island. Most people who visit the town pay tribute to the exceptional friendliness and courtesy of the inhabitants, who quite often extend invitations to visitors to attend one of their *ceilidhs* (smoking, drinking, and dancing concerts), a great feature of life on Barra.

It is worth while climbing Heaval to enjoy the splendid views across the other Barra Isles. About halfway up the hill on the SE. side is a statue of the Virgin Mary.

Barra: Castle Bay (Aerofilms)

If you continue on to the N., round the W. side of Hartval to the 550-ft pass, and then turn W. for about a mile. you will find a chambered cairn situated on a hillock.

From here one can travel on to Borve Point, where there is an ancient burial ground well worth seeing. There are other hills, Ben Tangaval (1,092 ft) being the chief. It is the variety of scenery which is the charm of Barra: the precipitous cliffs with their bird-life, the lochs, the remains of the Viking fortress of Dun Banm, the heather moors, and the delightful miniature valleys. There are four lochs and all of them provide excellent trout fishing.

At Northbay, the other important settlement on Barra, is the Church of St Barr, after whom Barra is named. Little is known about St Barr other than legends handed down by word of mouth. It is said that when he arrived on Barra he found that people were not merely heathens but cannibals, and that they had just eaten the missionary priest who had preceded him there. He is said to have been a disciple of St Columba and to have built the first church on the island.

See also Barra Isles, Calvay.

Accommodation: there is a hotel in Castlebay, and many of the islanders take in visitors.

Post Offices: at Castlebay and Northbay.

Books: *The Enchanted Isles*, by Alasdair Alpin MacGregor, Michael Joseph, 1967; *Rambles in the Hebrides*, by Roger A. Redfern, Hale, 1966.

BARRA ISLES

SITUATION: off the southernmost tip of the Outer Hebrides, 5 miles SW. of South Uist, 108 miles from Oban, the chief Scottish mainland port.

TOTAL AREA: 35 sq. miles.

POPULATION: 2,250 in all, but most islands are uninhabited.

ACCESS: by boat from Oban to Castlebay on Barra Island, Mondays, Wednesdays, and Fridays, leaving Oban at 7.00 am (the voyage takes 10 hrs); or by air from Glasgow daily.

Some fifteen of the fifty-odd Barra Isles are inhabited, but the largest island and that with most inhabitants is Barra (q.v.) itself. Of the others the most interesting are Fuday (q.v.), Fuiary (q.v.), Flodday, Gighay, Hellisay (q.v.), Kisimul (q.v.), and Huldoanich, but nearly all the isles have a special appeal for the ornithologist, archaeologist, and geologist. There is much to see, from old castles and ancient buildings to delightfully wild and verdant heathland, with rare flowers sometimes to be found in the most unexpected places.

The threat of depopulation has caused some concern

in the islands for many years. In 1958 Mr Robert Lister Macneil, the American chief of the Clan Macneil of Barra, sold some 9,000 acres of the islands, including all the outer inhabited isles. Barra is the ancient stronghold of the Clan Macneil, and many of the crofters on the islands bear the name. Clan Macneil, famed for its bards and pipers, came to the isles in the 11th century. The present clan chief is forty-fifth of his line, and in direct descent from the Irish King Niall of the Nine Hostages. He was born in the USA, where he was an architect in Vermont until he returned to his heritage in recent years.

The tradition of the islands is still mainly Roman Catholic and (though now only among the older inhabitants) Jacobite. Some of them still drink a toast to 'The King over the water', as they call the Stuart heir. They have a taste for fantasy and are believers in ghosts and fairies. Their chief recreation is the *ceilidh*, a smoking concert at which whisky is consumed in considerable quantities. On Barra itself they often sing and dance into the early hours of the morning. They will tell you that ghosts are always green in colour, but one suspects this may be the result of a post-*ceilidh* hangover!

See also Barra, Fuday, Fuiary, Hellisay, Kisimul.

Books: *The Enchanted Isles*, by Alasdair Alpin MacGregor, Michael Joseph, 1967; *Summer Days among the Western Isles*, by the same author, Thomas Nelson, 1929.

BASS ROCK

SITUATION: 3 miles off Canty Bay, North Berwick, East Lothian.
AREA: 1 mile in circumference.
POPULATION: 4.
ACCESS: by boat trips from North Berwick.

Bass Rock, Firth of Forth (Scottish Tourist Board)

This huge, precipitous mass of rock rises 350 ft out of the sea and is today the home of sea birds, principally gannets of a species known to ornithologists as *Sula bassana*, taking their name from the Rock.

The rock is comprised of hard, igneous stone, a relic of ancient volcanic masses. One of its most remarkable features is a tunnel which passes right through it from E. to W., caused by a natural fissure in the rock.

Bleak and forbidding as it may seem today, the Bass Rock has a rich history, dating back to the 7th century, when St Baldred had a cell on the island. After this it became a centre of the Celtic Columban Church and later, because of its impregnable position (there is only one landing place in its otherwise precipitous coastline), it became a fortress. During the reign of Charles II it was used as a prison for Covenanting ministers. But the most noteworthy incident in its history was in 1691, when a group of Jacobites captured the Rock and claimed it for the abdicated monarch, James II. They remained there for three years.

It has been said that the Bass Rock was used as a prison after the Rebellion of 1745, but this is merely a fiction employed by R. L. Stevenson, who introduces the island into his book *Catriona* as the place where David Balfour was confined.

The population of four today is comprised entirely of the lighthouse keepers. Application has to be made for permission to land, and this is sometimes granted locally to genuine bird-watchers.

BENBECULA

SITUATION: midway between North Uist and South Uist in the Outer Hebrides, about 1 mile from each and 22 miles W. of Skye.

AREA: 6 miles by 5 miles.

POPULATION: 1,182.

ACCESS: by boat from North or South Uist or, conditions permitting, by causeways linking the island with the Uists. Also by air from Glasgow, Aberdeen or Inverness to the airfield on Benbecula.

Despite its two-way causeway link with the Uists, Benbecula is in many respects an island in its own right, especially as it has its own air strip. The E. side of the island is a maze of tiny isles, too small to mention in detail, and sea lochs, while on the W. there are flat, fairly fertile plains, protected from the sea by miles of sand dunes.

It is a low-lying island with one hill rising to 409 ft above sea level – Rueval, where Prince Charles Edward Stuart sought refuge after the 1745 Rebellion. From the summit Benbecula seems to be an island of lochs, for the interior is sprinkled with them to such an extent that on a sunny day they dazzle the eyes as one looks down on this sparkling crystal landscape.

The chief community on Benbecula is at Balivanich, where the airfield is situated, at the NW. corner of the island. Angling is available on the island, as one would guess with so many lochs, and the sand dunes, surprisingly enough, are sufficiently sheltered to make picnicking a delight.

Separating Benbecula and South Uist is the South Ford, a shallow strait where an old track crosses the exposed sands at low water. Nevertheless it is dangerous to make this crossing unless one has judged the state of the tide correctly, and many people have been drowned attempting it.

Accommodation: a limited amount is available.

BERISAY

SITUATION: close inshore to Bernera, W. of Lewis in the Outer Hebrides.
AREA: 28 acres.
POPULATION: uninhabited.
ACCESS: by boat from Bernera.

This island's name and its location are apt to be confusing. Berisay is sometimes known as Birsay, and must not be mistaken for the Orkney township of that name. Similarly, its neighbour Bernera must not be confused with Berneray (q.v.), an island further S. in the Outer Hebrides.

It was on Berisay that Neil MacLeod, illegitimate son of 'King' Roderick of Lewis, rallied his followers against Mackenzie of Kintail and his forces. Though outnumbered, he so constantly harried the enemy that eventually they decided to leave the island. Neil MacLeod remained, but led a life which was little better than piracy. He was eventually taken prisoner and executed in Edinburgh.

Incidentally Bernera, mentioned above, disqualified itself as an island in 1953 when it was linked to Lewis by a bridge for which its inhabitants had been campaigning for some years.

BERNERA: *see under* Berisay.

BERNERAY

SITUATION: the most southerly of the Outer Hebrides, 1 mile S. of Mingulay.
AREA: 446 acres.
POPULATION: 3.
ACCESS: by boat from Mingulay or Barra.

Easily identified from the sea by its lighthouse on Barra Head, towering 683 ft above high-water mark, Berneray has a precipitous coastline of cliffs extending for six miles. The light can be seen on a clear night from a distance of nearly 40 miles.

Close to the lighthouse is an enormous chasm measuring 300 ft by 550 ft, which leads to a cave the dimensions of which are said not yet to have been established: surely a challenge for some keen spelaeologist!

As one would expect in the vicinity of such magnificent cliffs, Berneray is thronging with bird life, but being one of the least accessible of the Outer Hebrides, it is not often visited by bird-watchers. There is some grazing land on the island and the three lighthouse keepers take it in turns to mind the sheep. A century ago the island had more than thirty inhabitants. When it was offered for sale in 1951 the advertisement stated: 'there are no rates . . . there is shooting in the autumn – but no fishing other than in the sea, where lobsters are plentiful. Basking sharks, huge but harmless, are common at some seasons . . . 200 sheep.'

The island was not disposed of in 1951, but three years later it, together with Mingulay (q.v.) and Pabbay (q.v.), was sold to a partnership of five residents of Barra known as the Barrahead Sheep Stock Company.

BIGGA

SITUATION: in Yell Sound, 1 mile W. of Yell and 1 mile E. of Shetland Mainland.
AREA: 235 acres.
POPULATION: uninhabited.
ACCESS: by boat from Yell or Shetland Mainland.

In 1931 Bigga was listed as having one inhabitant. However, it is probable that in previous years, when census returns for Bigga gave the population as nil, there were several more people living there, as it was supposed to be common to the parishes of Delting and Yell. Certainly there were more than twenty people living there in the early part of the last century.

Bigga has good grazing land and offers a variety of scenery, wild life, and Shetland plants to the nature lover.

See also Shetland Isles.

BIRSAY: *see* Berisay.

BOGHA CORR: *see under* Sulasgeir.

BOGHA LEATHAINN: *see under* Sulasgeir.

BOGHANNAN S'IAR: *see under* Sulasgeir.

BORERAY (St Kilda Isles)

SITUATION: one of the St Kilda group of islands, lying 3½ miles NE. of Hirta and about 36 miles WNW. of North Uist.
AREA: 189 acres.
POPULATION: uninhabited.
ACCESS: by boat from Hirta.

Stark, black, precipitous cliffs towering out of the sea typify Boreray, which rises to a height of 1,245 ft. The cliffs range from 300 ft to 1,000 ft and because of them access to the island is difficult at all times of the year. There is one landing site at the south end of the island.

Yet there are many indications of ancient habitations on the island. There is a ruined structure known as Staller's House, believed to have been built by one Staller who seized the island from the MacLeod clan and built himself a miniature fortress here. In the middle of the 18th century the Revd Kenneth MacAulay described Staller's House as being

> 18 feet high, and its top lies almost level with the earth, by which it is surrounded; below it is of circular form, and all its parts are contrived so that a single stone covers the top. . . . In the middle of the floor is a large hearth; round the wall is a paved seat on which 6 persons may conveniently sit. There are 4 roofed beds roofed with strong flags or strong lintels, every one of which is capable to receive 4 men.

Staller's House is said to have extended underground, so that it could provide protection in the event of the island being besieged, and to have had a passage leading to a sea cave 700 ft below.

One of the most interesting discoveries on Boreray was made comparatively recently. Various tests and checks were made of surveys of megalithic sites all over the British Isles, and from these was developed the theory that the summit of Boreray was in ancient times used as a calendar marker linked to *menhirs* (stones) built about 1790 BC on the Outer Hebrides. It has been found that with the sun setting behind Boreray all these stones give calendar readings.

Close to Boreray are two barren rock islets that hardly qualify for the title of island, though one of them, Stac Lee (area 6 acres) has been described by Sir Julian Huxley as 'the most majestic sea rock in existence'. Stac Lee rises to a height of 544 ft and lies about 600 yds W. of Boreray. The other rock, Stac an Armin, is just under ¼ mile NW. of Boreray, somewhat larger than Stac Lee, and 627 ft in height. Both these rocks and Boreray are nesting grounds for gannets.

See also St Kilda Isles.

BORERAY (Sound of Harris Islands): *see under* Sound of Harris Islands.

BOTTLE ISLAND

SITUATION: 2 miles E. of Priest Island in the Summer Isles, and 3½ miles NW. of the Ross and Cromarty mainland.
AREA: 90 acres.
POPULATION: uninhabited.
ACCESS: by boat from Ullapool.

Bottle Island is so called because from a distance it resembles three bottles bobbing about in the sea. There are in fact three tiny islands lying close together.

See also Summer Isles.

BOUND SKERRY: *see under* Skerries (Shetlands), the.

BRESSAY

SITUATION: ¾ mile E. of Shetland Mainland.
AREA: 11 sq. miles.
POPULATION: 250.
ACCESS: by boat from Lerwick, Shetland Mainland.

Bressay, which is 6 miles long from N. to S., is one of the most fertile isles in the Shetlands (q.v.). It gives a natural protection to Lerwick harbour, and indeed the whole of Bressay Sound, which separates the island from Shetland Mainland (q.v.), is an admirable shelter for shipping. At one time, when the Dutch fishing fleet came here in the 15th century, the Sound was filled with shipping and Bressay became its headquarters. In the 17th century, it is recorded, as many as 1,500 ships anchored in the Sound.

The island rises to 743 ft at the conically-shaped Ward Hill in its S. area, and here also are the headlands of the Ord and the Bard. One of the chief attractions of the island is the Cave of the Bard, a stalactite cave on the W. side of Bard Head (264 ft) at the S. extremity of Bressay, noted for its pronounced and reverberating echo.

There is more farmland for its size on Bressay than on most other Shetland islands: most of the produce is sold across the water at Lerwick, with which port the island links its fishing industry, having a fish-meal factory at Heogan. The population is concentrated mainly on the W. side, facing Lerwick, though there are a number of crofts on the E. side.

In 1665 Bressay Sound was the scene of a successful foray by the English fleet against the Dutch. A tombstone was erected on Bressay in 1636 in memory of a Dutch naval commander who died there. This tombstone, which can be seen today in the ruins of St Mary's Church, close to the site of an ancient *broch* at Cullingsburgh in the NE., is inscribed in Dutch. The translation reads: 'Here lies buried the brave commander, Claes Jansen Bruyn, of Durgendam, who died in the service of the Dutch East India Company, August 27, 1635.'

Shetland ponies were bred on Bressay for many generations, the mares being kept here and the stallions on Noss (q.v.). A former Marquis of Londonderry kept a pony farm here many years ago to supply ponies for his coalmines. John Tudor writes that at a sale held in 1878, 'thirty lots of these "horse-ponies" realised an average of £25 apiece'.

For the visitor Bressay offers an excellent golf course, and much scope for the ornithologist. The eiderduck nests here, and in June one can come across many of its nests. What are known locally as the 'plantie-crubs' – rectangular drystone enclosures – provide nesting-places for a variety of birds: starlings, wrens, blackbirds, and others. Apart from this the 'plantie-crubs' often afford shelter for a wide variety of plant life – tiger-lilies, lupins, michaelmas daisies, rose-root, and the somewhat bizarre Australian daisy.

There are magnificent views across the whole of the Shetlands from Ward Hill, taking in Yell (q.v.), Fetlar (q.v.), and Unst (q.v.) to the north, and even far-distant Fair Isle (q.v.) on a good day.

Accommodation: limited, but many of the cottagers take in visitors.

BROTHER ISLE

SITUATION: in Yell Sound between Yell and Shetland Mainland.
AREA: 126 acres.
POPULATION: uninhabited.
ACCESS: by boat from Shetland Mainland.

BRURAY: *see under* Skerries (Shetlands), the.

BURRA

SITUATION: 1 mile W. of Shetland Mainland.
AREA: 5½ miles from N. to S. and 1–1½ miles wide.
POPULATION: 663.
ACCESS: by boat from Scalloway, Shetland Mainland.

With its splendid natural harbour at Hamnavoe, Burra is an ideal centre for the fishing industry in this corner of the Shetlands (q.v.), especially as there is an abundance of fish in the adjacent waters. In the past fishing has brought a modest measure of prosperity to Burra, which is reflected in the well-built cottages, nearly all of which have water and electricity. Nevertheless the population has been dwindling for some time.

Burra is really comprised of two islands, West and East, but the two are linked by a bridge which has the effect of making this an H-shaped island. It is a pleasant, green landscape, with some splendid walks and even more splendid views, not only of Shetland Mainland (q.v.), but of several other islands in the vicinity, Trondra (q.v.) and several small islets to the NW.

Papil in West Burra is reputed to have had a tall church

once, rather like that on Egilsay (q.v.) in Orkney. Whether or not this is true, the Papil Stones, fine examples of the early Celtic cross, were found here. One of the Stones still remains here, but the other is in the National Museum in Edinburgh.

Accommodation: some cottagers occasionally take in visitors.

BURRAY

SITUATION: ½ mile W. of Orkney Mainland.
AREA: 1 mile by ½ mile.
POPULATION: 30.
ACCESS: by road from Kirkwall.

Burray was entirely an island until World War II, when it was linked to Orkney Mainland (q.v.) on the creation of what became known as the Four Churchill Barriers. These were built by some 550 Italian prisoners-of-war captured during General Wavell's offensive in North Africa. They had been taken to Orkney, and in 1943 helped to erect these barriers across the eastern approaches to Scapa Flow.

Eventually the barriers were used to link Burray and South Ronaldsay (q.v.) to the Mainland of Orkney. They have saved Burray from what otherwise would have been total depopulation, for today there is a busy boat-building yard on the island. More than 400 boats ranging from dinghies to 36-footers have been built here since the war, and the industry is still expanding.

See also Orkney Islands.

BUTE

SITUATION: separated from the Argyll mainland by a very narrow channel, the Kyles of Bute, in parts only ¼ mile wide, on its N., NW., and NE. sides.

AREA: 47 sq. miles.

POPULATION: 19,465.

ACCESS: by regular boat services to Rothesay from Wemyss Bay and Gourock (there are car ferry services).

Though smaller than the Isle of Arran near by, Bute is nevertheless a county in its own right, as are Orkney and Shetland. As its northern tip fits so neatly into the contours of Argyll, from which it is separated only by the narrowest stretch of water, in many respects it seems less like an island than most. But it has some beautiful scenery and some of the most fertile land in the W. of Scotland, and for many years it has been a favourite resort with holiday-makers.

Its principal town and port is Rothesay, which has many excellent hotels and some splendid beaches. It was made famous by Mrs Craik with the song which she wrote and set to music, 'Sweet Rothesay Bay'. It has a winter garden, three golf courses, bowling greens, tennis courts, a concert hall, and cinemas.

Robert III of Scotland made Rothesay a royal borough and created the title of Duke of Rothesay for his son. Today this is one of the titles of the Prince of Wales. Rothesay Castle, founded in the 11th century during the Norse occupation by Magnus Barfod of Norway, is one of the outstanding medieval castles of Scotland. The existing remains, however, date mainly from the 14th century. Cromwell set about destroying it in 1650, but much remains that is of interest. It is circular in plan, with high curtain walls, and was originally flanked by

opposite: *Bute: a neolithic cairn (British Tourist Authority)*

four round towers enclosing a circular courtyard. The whole site is surrounded by a deep water-moat, across which projects a high tower which formed the entrance.

The castle is in the old town of Rothesay, about 3 minutes' walk from the pier. Near by is the ruined chapel of St Michael. Beside the parish church of Rothesay is the ruined chancel of the old Gothic church of St Mary, especially interesting for its canopied tombs with effigies.

Many pleasant excursions can be made into the interior of Bute. There is, for example, Barone Hill (530 ft), which gives an admirable view of the group of lochs in the interior of the island S. of Rothesay. On the W. side of one of these lochs, Loch Fad, is 'Kean's Cottage', a Regency-style house built by the actor Edmund Kean in 1827. This house was later occupied by Sheridan Knowles, author of *The Hunchback* and *The Love Chase*. Another worthwhile trip is to Kames Bay where stands the tower of Kames Castle, said to date from the 14th century.

Unfortunately some of the finest bays on the island, with splendid stretches of sand, such as Ettrick Bay on the W. coast, have been spoilt by tourist and tripper crowd attractions. At Kilmichael, 4 miles N. of Ettrick Bay, are the ruins of an ancient chapel. A rewarding trip is that to Kilchattan Bay, some 9 miles S. of Rothesay, taking in the model village of Kerrycroy and Mount Stuart, the ornate Victorian residence and seat of the Marquesses of Bute, who are descended from the hereditary Stuart Keepers of Rothesay Castle. This was built in 1877, and is set in one of the more delightful corners of the island. Not much further S. is the site of St Blane's Church, a Culdee foundation dating from the days of the earliest Christian missionaries. The road here skirts the Standing Stones of Lubas, and the ruined nave and chancel of St Blane's edifice can still be seen. N. of the ruins are remains of a circular structure about 32 ft wide known as the Devil's Cauldron.

The highest point on the island is 911 ft at Kilmichael in the N., and to the W. of this is the beautiful Glen

More. Bute is a strange mixture of the old and the new, the solitary and the suburban. Sometimes it seems merely a holiday area for Glaswegians, yet it is possible to escape to remote places that capture the island atmosphere. It was the first place in Scotland in which a cotton mill was built – in 1788 by David Dale of Glasgow – and a cotton industry is still maintained on the island. But apart from tourism, the chief occupation of islanders is farming – mainly dairy farming. The Scottish Milk Marketing Board set up a creamery in Rothesay in 1954.

Accommodation: there is a wide selection of hotels of all classes in Rothesay and in other parts of the island.

Post Offices and banks: facilities are extensive.

Books: *The Inner Hebrides and Their Islands*, by O. F. Swire, Collins, 1964; *Scottish Castles: an Introduction to the Castles of Scotland*, by W. D. Simpson, HMSO, 1959.

CALF OF EDAY: *see under* Eday.

CALVAY

SITUATION: in the Sound of Eriskay, close to Eriskay in the Outer Hebrides and due S. of South Uist.

AREA: 10 acres.

POPULATION: uninhabited.

ACCESS: by boat from Eriskay.

Too often the claim has been made that Sir Compton Mackenzie's novel, *Whisky Galore*, was based on a wreck that occurred on Eriskay or Barra. As both the latter islands have more serious claims to historical consideration it is only fair to give the insignificant islet of Calvay credit for what actually took place.

In 1944 the British merchant ship *Politician*, of 12,000

tons, on her way to America with a cargo of the best Scotch whisky, was torpedoed by the Germans and went ashore on Calvay. Thousands of cases of whisky were retrieved from the ship before she went down by intrepid islanders making trips out to the wreck from Eriskay and Barra. The whisky thus rescued was known locally as 'Polly', a contraction of *Politician*, and for a time it was so plentiful in a period of acute whisky shortage that housewives actually used bottles of it to help kindle stubborn fires, while their menfolk, apart from drinking the stuff, found it ideal for cleaning oil and grease from their hands.

Out of this incident Sir Compton Mackenzie created *Whisky Galore*, since made into a film, a story which captures in the most authentic fashion the characters of many islanders, and especially of the Roman Catholic community who predominate in these parts. In Mackenzie's book the *Politician* becomes the SS *Cabinet Minister*. John Macpherson of Barra, locally known as 'The Coddie', played a part in the film, made by Ealing Studios, and filmed on Barra over a period of twelve weeks. He was shown distributing the whisky at a *reiteach* – the traditional drinking festivity which precedes a wedding in the islands.

For the record, Eriskay has been named as the real site of *Whisky Galore* because it was the nearest large island to Calvay. Barra has been named partly because the film was made on the island and partly because in 1940 the SS *Jamaica Progress* was torpedoed a mile offshore. She was carrying a cargo of bananas and puncheons of overproof spirit of rum. In the mid-fifties storms broke up the wreck of the *Jamaica Progress* and released some of her cargo on the shores of Barra.

CANNA

SITUATION: in the Small Isles group of the Inner Hebrides, $2\frac{1}{2}$ miles NW. of Rhum.
AREA: $4\frac{1}{2}$ miles by 1 mile.
POPULATION: 30 (1970).
ACCESS: by steamer from Mallaig.

Canna is sometimes referred to as the 'Garden of the Hebrides'. It is a fertile island which, as it lies E. and W., is protected from the chill N. and NE. winds, and in consequence many plants and crops grow here which would not normally be found in the Hebrides. An unusual feature is the large variety of moths to be found here in summer.

Its highest point is some 690 ft above sea level, and there is a good (if small) harbour, sheltered by the miniature island of Sanday, to which Canna is connected at low tide. Canna is the only one of the Small Isles (q.v.) in which a steamer can tie up. Compass Hill on the N. of Canna is so named because its magnetic rocks have been known to affect the compasses of passing ships. Many fishing vessels and small craft put in at the harbour. The coastline is mainly rocky, but interspersed with a few sandy bays and a number of caves.

The inhabitants are mainly engaged in crofting and lobster fishing. The soil is a reddish clay. At Tarbert there is a deserted homestead, and on the S. side below Tarbert are the remains of an ancient holy building, probably a nunnery, as it is called Sgeir nam Nan Naomha, 'the Skerry of the Holy Women'. At the E. end of Canna stands a pillar of rock 100 ft high which is joined to the island by a narrow stretch of rock, on the summit of which is the ruin of Coroghon Mor, a prison built by a chieftain of Clan Ranald, in which he incarcerated his wife to keep her safe from possible marauders. A notice advises against attempts to climb the rock.

Unlike its larger neighbour, Rhum (q.v.), Canna has

preserved its population and traditional way of life in this century, thanks mainly to its Gaelic-speaking laird and owner, Dr John Lorne Campbell who, with his American wife, who also speaks Gaelic, has helped the island to retain its economic independence and preserve its native culture. Both Dr Campbell and his wife are Gaelic scholars who have made intensive studies of the history of Canna.

There is a Roman Catholic church for the islanders, and also a Protestant church which is mainly used by visitors.

Accommodation: there is no hotel on Canna, but some cottagers take visitors, and one such is situated close to the approach to Sanday Island.

CARA

SITUATION: 1 mile S. of Gigha Island, Kintyre.
AREA: 1¼ miles by ¼ mile.
POPULATION: uninhabited.
ACCESS: by boat from Ardminish in Gigha.

Unlike its neighbouring islands, Gigha (q.v.) and Gigalum (q.v.), Cara is owned by the Macdonalds of Kintyre, who have held it since about the 14th century.

A ruined house and a chapel survive to indicate the island's previous occupation. The feature of the island which is most conspicuous is the 'Brownie's Chair', a prominent stone shaped like an armchair. 'The Brownie' is supposed to be the fairy protector of the Macdonalds, the ghost of a Macdonald murdered by a Campbell centuries ago.

Permission should be sought before landing on this island.

Book: *The Antiquities of Gigha*, by R. S. G. Anderson, Galloway Gazette Ltd, 1936.

CARN BEAG: *see under* Carn Isles.

CARN DEAS: *see under* Carn Isles.

CARN IAR: *see under* Carn Isles.

CARN ISLES

SITUATION: 5 miles W. of the Ross and Cromarty mainland, and 16 miles NW. of Ullapool.
TOTAL AREA: 50 acres.
POPULATION: uninhabited.
ACCESS: by boat from the mainland.

This is a group of some five islets within the larger area of the Summer Isles (q.v.). They are Carn Iar, Carn Beag, and Carn Deas, and two knoll-like islands, East and West Carn. The larger islands are joined by a shingle causeway which is passable at all tides.

They are used solely for sheep grazing by crofters from Achiltibuie, and possess no buildings or even ruins. A small colony of shags and a few fulmars are to be found on Carn Deas. Vegetation is mainly heather, with very little grassland. All the islands were up for sale some few years ago, the prices then ranging from £1,850 for the smallest to £5,250 for the largest, but it is doubtful if these prices would obtain today.

CASTLE ISLAND: *see under* Cumbraes, the.

CAVA

SITUATION: 2 miles E. of Hoy and 4 miles S. of Orkney Mainland.
AREA: 180 acres.
POPULATION: uninhabited.
ACCESS: by boat from Lyness or Orphir.

Now used solely for occasional grazing purposes, Cava is interesting because it contains the remains of one of the typical ancient 'Orkney houses' – long houses with barns attached and the ingenious built-in, hollowed-out furniture of flagstone.

See also Orkney Islands.

CEANNAMHOR: *see under* Badcall.

CHEYNIES

SITUATION: 2 miles W. of Scalloway, Shetland Mainland.
AREA: 54 acres.
POPULATION: uninhabited.
ACCESS: by boat from Scalloway.

CIARN A'BURG

SITUATION: ¾ mile S. of Fladda and 2 miles NE. of Lunga in the Treshnish Isles, 4½ miles from Mull in the Inner Hebrides.
AREA: 35 acres.
POPULATION: uninhabited.
ACCESS: by boat from Lunga or Mull.

These two islets, Ciarn A'Burg Mhor and Ciarn A'Burg Beg, are the most prominent of a score or more of rocks and skerries lying in the sea around Fladda. They are flat-topped, like most of the Treshnish Isles (q.v.), and the only items of interest are the remains of a ruined castle on the Ciarn A'Burg Mhor and a small fort on the Ciarn A'Burg Beg.

COLL

SITUATION: 10 miles WSW. of Ardnamurchan in Inverness-shire and 7 miles W. of Mull in the Inner Hebrides.
AREA: 28¾ sq. miles.
POPULATION: 140 (1970).
ACCESS: by steamer from Oban to Arinangour.

Lying between the islands of Mull and Tiree, from which last it is separated by just over 2 miles of water, Coll is justly famed for its trout-filled lochs. It never rises above about 340 ft, and has a grassy and arable interior but a grimly rocky east coast.

Its history dates back to ancient times, and it was at one time occupied by the Norsemen. In Coll's many lochs are to be found 'duns', small forts surrounded entirely by water, to which the inhabitants would retire when the island was invaded.

Coll's population once exceeded 2,000, but once this peak was reached, about the beginning of the 19th century, crofting, the island's chief occupation, began to decline. Later in the century many crofters were evicted. Yet there are many far smaller islands in the Inner Hebrides with less productive land than Coll which support a larger population today.

The chief village and port is Arinangour, half-way down the E. coast. The western side of the island is more attractive, having several bays of golden sand. At

Breacacha at the S. tip of Coll stand the ruins of a fortress of the Macleans, formerly the lairds of Coll. Close to this fortress is a monstrous 18th-century house built by the Macleans when the fortress became uninhabitable. Dr Johnson on his visit to Coll referred to it contemptuously as 'a tradesman's box'.

The island will appeal to all trout fishermen. Anglers say that the brown trout of Coll are 'lively, a challenge and never too small to be a disappointment'.

Accommodation: there is a hotel at Arinangour, and some cottagers take visitors for bed and breakfast. There are also scheduled camping sites on the island.

COLONSAY

SITUATION: in the Inner Hebrides, 10 miles W. of the Isle of Jura and linked at low tide to the neighbouring island of Oronsay.

AREA: 16 sq. miles.

POPULATION: 275.

ACCESS: by boat from West Loch Tarbert, or from Port Askaig on Islay.

An attractive island of craggy hills, woods, and a rocky coastline with silver sands and high cliffs, Colonsay offers much of interest to the botanist, archaeologist, and ornithologist. Gannets and fulmars are to be found everywhere, and grey seals breed here.

Colonsay has a pier, and is linked with Tarbert on the Argyll mainland by steamer service. Its principal settlement is at Scalasaig on the E. side of the island. Lord Strathcona is the owner of Colonsay, and before you can camp on the island permission must be obtained from him. There is little difficulty in getting this permission, the only stipulations being that this is not to be taken as establishing a precedent for future years, and that you camp at Machrins Bay.

Colonsay House, at the N. end of the island, is the home of Lord and Lady Strathcona. Its typical Hebridean gardens contain some sub-tropical plants, and under the Scottish Garden Scheme they are open to the public in summer. There are palm trees as well as an abundance of rhododendrons and azaleas. This is the most attractive part of the island – the sheltered Vale of Kiloran – and near by is a magnificent beach known as Kiloran Sands, the main occupants of which are cows who bask there in the sunshine.

Accommodation: the island has a hotel, and several cottages offer bed and breakfast.

COLSAY

SITUATION: 2 miles W. of Scousburgh in the S. of Shetland Mainland.
AREA: 54 acres.
POPULATION: uninhabited.
ACCESS: by boat from Scousburgh.

COPINSAY

SITUATION: 2 miles SE. of Point of Ayre on Orkney Mainland.
AREA: 200 acres.
POPULATION: uninhabited.
ACCESS: by boat from Skaill.

Once this island was tenanted and had a farm-house and chapel. Today it is virtually deserted except by the lighthouse attendant and by the gulls, kittiwakes, razorbills, shags, cormorants, guillemots, and fulmars which come here in summer.

In 1972 Copinsay was bought by nature-lovers for £7,500 as a memorial to James Fisher, the ornithologist and broadcaster who was killed in a car crash in 1970. The appeal to raise funds to buy the island was sponsored by some eighteen national and local organisations, headed by the World Wildlife Fund. A count on Copinsay recently recorded more than 10,000 kittiwakes, 9,000 guillemots, and 700 fulmars on the island. Copinsay is now managed by the RSPB.

The island has sharply-rising cliffs, reaching 200 ft on its E. coast, close to which pass the Aberdeen ships travelling to Orkney. The W. side of the island is in strong contrast, with its green, now untended slopes. A tall, curiously-shaped stack of rock, topped with green turf, rises on the NE. side of Copinsay, and almost constitutes a separate island. It is known as the Horse of Copinsay, and is said to have been used for sheep-grazing at one time. Certainly its forbidding slopes would deter all but sheep and goats, and it may be that their grazing here gave rise to the local legend that on the Horse of Copinsay you could fatten one sheep, feed two and starve three.

See also Orkney Islands.

CRAIGLEITH

SITUATION: 2½ miles W. of the Bass Rock in the Firth of Forth, and 1 mile N. of North Berwick, East Lothian.

AREA: 20 acres.

POPULATION: uninhabited.

ACCESS: by boat from North Berwick.

A rocky, barren isle with some bird life, Craigleith is

also known as Lamb Island. It rises to a height of 168 ft, and at low water is separated from the coast by a channel half a mile wide with a depth of 6 fathoms in the fairway.

This island should not be confused with Lamb Islet, 1¼ miles WNW. of North Berwick, which is 79 ft high and nothing more than a totally barren rock surrounded by foul ground on its NE. side.

CRAMOND

SITUATION: in the Firth of Forth 1 mile N. of Cramond, Midlothian, and 5 miles WNW. of Leith.
AREA: 80 acres.
POPULATION: 7.
ACCESS: by boat from the village of Cramond.

The island rises to a height of 91 ft.

CREAG A'CHASTEAL ISLAND: *see under* Lunga.

CROIS SHITINIS: *see under* Monach Isles.

CROWLIN ISLES

SITUATION: 3 miles SSE. of Toscaig, Inverness-shire mainland, 7 miles S. of Applecross, and 9 miles E. of Skye.
TOTAL AREA: 420 acres.
POPULATION: 7.
ACCESS: by boat from Toscaig.

This is a group of three islands connected with one another by reefs that dry out at low water. Eilean Mor, the largest, is 367 ft high, and its E. and S. sides are fringed by reefs which extend in places about ½ cable offshore. The reefs are teeming with much interesting marine life and plants.

Eilean Meadhonach lies close to the W. of Eilean Mor, and Eilean Beag, the most northerly of the group, has a light beacon.

THE CUMBRAES

SITUATION: in Millport Bay, Bute County.
AREA: Great Cumbrae is 3¾ miles by 2 miles. Little Cumbrae is less than 2 miles by 1 mile.
POPULATION: Great Cumbrae 5,934; Little Cumbrae 18.
ACCESS: by boat from Largs on the Scottish mainland, or from Rothesay on Bute.

The Cumbraes (Great Cumbrae, Little Cumbrae and the islets known as the Eileans) are equally delightful for a day's visit or for a holiday, and Millport in Great Cumbrae has been developed as a resort. It has good sandy beaches admirable for bathing, a harbour with a ferry service to Largs, and a golf course.

A 12-mile road runs right round Great Cumbrae, the northern island, and this takes in Keppel, where the Scottish Marine Biological Station is established, with a museum and aquarium which are open to the public. This is well worth a visit. Scientific research here covers a wide field, including the chemistry of the sea, an examination of prawns and their parasites, and the growth of barnacles. University students specialising in marine biology can attend classes here.

Millport has diversified its industries as well as concentrating on the tourist trade. It possesses a grain mill, a quarry, and an embroidery hand-loom factory where local women are employed.

There is an Episcopal Church in Millport, built in 1851, which in 1876 was made the 'Cathedral of the Isles'. It was not, however, the first church in the island: records show that in the 14th century the 'Chapel of Cumbrayne' was confirmed to Paisley Abbey by the Bishop of Glasgow, while in 1612 a new church dedicated to St Columba was built at Kirkton, ½ mile from Millport. Kirkton is believed to be the oldest habitation on Great Cumbrae.

There is a ridge of hills on Great Cumbrae which provides splendid views of the whole of the Firth of Clyde. Facing the promenade at Millport are the Eileans, tiny islets dotted around the bay, which beckon the lover of islands-in-miniature.

Little Cumbrae, to the S., separated from the larger island by half a mile of water, has less to offer the visitor, but its chief landmark is an ancient, ruined castle which was used by King Robert II as a royal residence – two of his charters were actually signed there. The Hunters of Hunterston were the hereditary keepers of the castle until 1515, after which it passed to the Earl of Eglinton and was eventually burnt by Cromwellian troops in 1653.

The only dwellings on Little Cumbrae are a farm and a lighthouse on the W. side. The latter was built in 1794, and a fog-horn was attached. An islet off the E. shore, Castle Island, has the ruins of a tower, also occupied by Robert II and destroyed by Cromwell. On the N. slope are the remains of a chapel dedicated to St Vey.

DA BOGHA LAMHA CLEIT: *see under* Sulasgeir.

DAVAAR ISLAND

SITUATION: 1 mile E. of Kintyre and 3 miles E. of Campbeltown.
AREA: 1½ miles by 1 mile.
POPULATION: 6.
ACCESS: by boat from Campbeltown.

Lying in the entrance to Campbeltown Loch, Davaar Island can on occasions be reached on foot across a spit of gravel at low tide. The island has a lighthouse, but its chief attraction is the cave on the SE. shore. On the rock wall of the cave Archibald McKinnon painted a crucifixion scene in 1887. The painting is visible in daylight because of a shaft of light which comes through a hole in the walls.

McKinnon touched up his painting some years later and, according to the latest reports, the scene was again touched up by a modern artist in the mid-1950s.

DHU HEARTACH

SITUATION: 20 miles SW. of Mull and 15 miles WNW. of Colonsay in the Inner Hebrides.
AREA: 10 acres.
POPULATION: uninhabited.
ACCESS: only by special arrangement with the lighthouse authorities.

Dhu Heartach is really a conglomeration of rocks on which is sited a lighthouse, built in 1867–72 from the red granite of Mull. The island is mentioned in a short piece entitled 'Memoir of an Islet' by Robert Louis Stevenson in his *Memories and Portraits*, in which he describes the building of the lighthouse.

DUN

SITUATION: about 120 yds from the southernmost point of Hirta, St Kilda Isles, and about 34 miles WNW. of North Uist.
AREA: 79 acres.
POPULATION: uninhabited.
ACCESS: by boat from Hirta.

Dun was once connected to Hirta (q.v.), but is today separated from it by a narrow channel in which many rocks jut out of the sea. It is about a mile long and has an average breadth of about 100–200 yds, yet rises sheer from the sea to a height of nearly 600 ft. Its highest peak is Bioda Mor, which is rocky on one side and covered with grass on the other. A landing on Dun is difficult and can best be made at low water. The St Kildans formerly

Dun, in the St Kilda Isles (Aerofilms)

used the island as a grazing area for lambs. There are remains of a hill fort on Dun (circa 9th century) which undoubtedly accounts for its name.

See also St Kilda Isles.

DUTCHMAN'S CAP

SITUATION: the most southerly of the Treshnish Isles, 2½ miles S. of Lunga, and 7 miles SW. of Mull in the Inner Hebrides.
AREA: 80 acres.
POPULATION: uninhabited.
ACCESS: extremely difficult, but possible from Mull or Lunga.

So called because of its resemblance to the traditional Dutchman's cap. It is the only one of the Treshnish Isles (q.v.) which is not flat-topped, being a rounded hump of rock, with scant vegetation, looming out of the sea.

EAGAMOL

SITUATION: close to the islet of Eilean Nan Each off the NW. tip of Muck in the Inner Hebrides.
AREA: 50 acres.
POPULATION: uninhabited.
ACCESS: by boat from Muck.

A haunt of seabirds, but also capable of supporting a few sheep in summertime owing to the lush grass produced by the natural fertilisers of bird droppings and dead fish.

EARRAID: *see* Erraid.

EASDALE

SITUATION: in the Inner Hebrides, less than $\frac{1}{4}$ mile W. of Seil Island, 9 miles SSW. of Oban.
AREA: 35 acres.
POPULATION: 83.
ACCESS: by boat from Seil.

Easdale was noted for its slate quarries which, until the sea broke in (1872), extended more than 200 ft below sea-level. Slate from here was used for re-roofing the abbey at Iona. The island was used for the production of material for making prismatic tools during World War II.

Accommodation: limited.
Post Office: in the village.

EAST BURRA: *see* Burra.

EAST CARN: *see under* Carn Isles.

EDAY

SITUATION: 3 miles W. of Stronsay island and 3 miles SE. of Westray, in the Orkneys.
AREA: $7\frac{1}{2}$ miles by $2\frac{1}{2}$ miles.
POPULATION: 474.
ACCESS: by boat from Stronsay or Orkney Mainland.

A hilly island of red sandstone, Eday is a stark and unfriendly-looking place when approached from the sea. It is from here that much of the peat which is cut and used as fuel in the Orkneys comes.

There are numerous weems and tumuli on the island, which is separated from the small Calf of Eday by the narrow Calf Sound, perhaps the most picturesque part of the steamer-route through the islands. The pirate John Gow was wrecked and captured here in 1725.

Like Shapinsay (q.v.), Eday owed much in the past to the foresight of its landlord, Robert Baikie, who, by giving his tenant a nineteen-year lease and thus security of tenure, encouraged him to enclose his lands, introduce ploughs, and improve his cattle and his buildings.

See also Orkney Islands.

EDDRACHILLIS ISLANDS

SITUATION: in Eddrachillis Bay, Sutherland, 1–3 miles off the coast.

TOTAL AREA: 3 sq. miles.

POPULATION: uninhabited, except for one or two islands where there is occasional seasonal occupation.

ACCESS: by boat from Scourie or Drumbeg.

There are some thirty-five islands in and around Eddrachillis Bay, all coming under the coastal parish of Eddrachillis. Two – Handa and Oldany – are fair-sized islands, but the rest are small and mainly insignificant.

Cultivable land is scarce here, as the terrain is of the same kind of formation of gneiss as is found in Lewis in the Outer Hebrides. For centuries the islands belonged to the Earls of Sutherland.

Handa, a nature reserve run by the RSPB, is the most interesting island, and also the largest.

See also Badcall, Handa, Oldany.

EGILSAY (Orkneys)

SITUATION: 2 miles E. of Rousay in the Orkneys.
AREA: 3 miles by 1 mile.
POPULATION: 45.
ACCESS: by boat from Rousay.

The name Egilsay is derived from the Gaelic word *eaglais*, meaning 'church', and obviously refers to the island's most notable landmark, St Magnus's Church. But some purists who insist that all Orkney place-names are derived from the Norse are adamant that the name comes from Egil, a Norse personal name.

This island is dagger-shaped, with the point directed to the S. Right in the centre of Egilsay on its highest point stands the Church of St Magnus, the round tower of which is visible for miles around – even from Orkney Mainland. What remains of this ancient church is impressive enough; its thick stone walls have stood the test of time and the rectangular nave, 30 ft long and 15 ft wide, still stands. No one has yet been able to date the church exactly, but it is believed to be 12th century. Probably it was founded by Bishop William the Old between 1135 and 1138, and named after Magnus, earl and saint, who was murdered on Egilsay in Easter Week, 1116. Magnus had frequently denounced the violence which the Norsemen practised in their raids and, though one of them, was also their sternest critic. By that time the Norsemen had embraced Christianity, and it was because of Magnus's saintly qualities that he was canonised in 1136 at the suggestion of Bishop William.

There is a small crofting community on Egilsay and a school with less than a dozen pupils. But there is little to see there except for St Magnus's Church.

See also Orkney Islands.

EGILSAY (Shetlands)

SITUATION: between Muckle Roe and Northmaven on Shetland Mainland.
AREA: 54 acres.
POPULATION: uninhabited.
ACCESS: by boat from Shetland Mainland.

EIGG

SITUATION: in the Small Isles group of the Inner Hebrides, 7½ miles from the W. coast of Inverness-shire and 6 miles SE. of Rhum.
AREA: 8½ sq. miles.
POPULATION: 120 (1970).
ACCESS: by boat from Mallaig.

A basaltic peak, the Sgurr of Eigg, rising to 1,289 ft, is the outstanding landmark of this island, which is rich in legend and ancient culture. Hugh Miller describes the Sgurr in his *Cruise of the Betsy* as 'a tower 300 ft in breadth by 470 ft in height, perched on the apex of a pyramid like a statue on a pedestal'.

Seen from the mainland on a sunlit evening, Eigg resembles a sinking ship at the moment its stern tips up before it finally disappears bows first. The island's name is pronounced 'Egg': it is an abbreviated form of Eilean Eige – 'the island of the indentation', which refers to the long depression or indentation which runs aross the centre of Eigg.

On landing at the small quay in the SE. corner the visitor makes his way along this indentation to Cleadale, the settlement of the crofters, now a dwindling community. There are innumerable walks across the island, which caters for a wide range of interests – those of rock-climbers, botanists, ornithologists, and geologists. The relatively few trees are to be found at the SE. end of Eigg, and there are said to be 379 different plants and flowers. The ridge of rocks that juts out from the Sgurr is a subject of controversy among geologists. It resembles, with its contorted and broken pillars of pitchstone, the columnar cliffs of Staffa. Even more fascinating in origin are the tiny lochs high up above the cliffs, where the Eigg vole and rare water insects are to be found.

The coast of Eigg has been compared by some to that of Iceland. The Sgurr, which dominates everything, rises out of the peaty moor and has precipices on three sides. Climbers should treat it with respect, for it is a formidable obstacle. It is best approached by starting from the far end on the N. and then walking back along the ridge. The boat trip normally allows three hours ashore, which gives just enough time for the climb.

It is legend rather than actual evidence of antiquity which typifies Eigg. The monks of St Columba established themselves on the island, and the inhabitants remained mainly Catholic. Irish Franciscan friars visited Eigg in the 17th century and maintained the Catholic tradition, and the Clan Ranald chiefs, who held sway here, continued to hold allegiance to Rome. But in time many Protestants came to live on Eigg, and there still exists on the island the site of a building which was once divided into three sections – the House of Devotion, the House of the People, and the House of the Sermon. This all-purpose church was built by the islanders for Catholics and Protestants to worship in. Today there are two small churches on Eigg – one Catholic, the other Church of Scotland.

In 1795 Eigg had 400 inhabitants. It has been repopulated mainly by the Macdonalds, and today has a single shop. Eigg has witnessed two major massacres in its history. According to the *Martyrology of Donegal*, in AD 618 'there came robbers of the sea on a certain time to the island, when he, Donan [St Donan, the ruins of whose chapel can still be seen] was celebrating mass. He requested of them not to kill him until he should have the mass said, and they gave him this respite, and he was afterwards beheaded and 52 of his monks along with him.'

These executions may have been ordered by the 'Amazon Queen' and her court of female warriors who lived in Moidart at this time and resented Donan. Legend has it that they coveted the island for themselves. Eigg was in fact sometimes referred to as Nim Ban More, 'the island of the big women'.

There is a cave near the NE. shore in which in the 16th century the MacLeods of Skye are said to have carried out another horrible massacre. They landed on the island to avenge a supposed wrong, lit a fire at the entrance to the cave, and thereby suffocated some 200 Macdonalds who were hiding there. Sir Walter Scott, after a visit to Eigg, said that he 'brought off a skull from among the numerous specimens of mortality which the cavern afforded.'

At the N. end of the island are the curiously named 'Singing Sands', which emit a strange humming and hissing sound when walked upon in dry weather. The best time to observe this phenomenon is said to be at sunset on a hot day. Bird-watchers should try to catch a glimpse of the golden eagle on Eigg – a bird which is nesting there again after a prolonged absence.

There are some attractive waterfalls on Eigg, some of which cascade down on to the edge of a raised beach. Of interest are a number of caves at the SE. corner of the island, including Cathedral Cave and Pigeon's Cave, where rock doves make their homes.

In 1966 the island, which had been in the possession of Viscount Runciman and his brother, Sir Steven Runciman, was put on the market at £100,000. Today it belongs to a trust which encourages self-help among handicapped boys.

See also Small Isles, Eilean Chathastail.

Accommodation: this is limited to cottages offering bed and breakfast. Visitors wishing to stay on the island should write

to The Factor, Eigg Estates Ltd, Isle of Eigg, Inverness-shire.
Books: *Cruise of the Betsy*, by Hugh Miller, 1958; *An Island Here and There*, by Alasdair Alpin MacGregor, 1972.

EILEACH AN NAOIMH

SITUATION: at the extreme SW. of the Garvellochs group of islands, close to the Isle of Luing and 5 miles W. of the Argyll coast at the entrance to the Firth of Lorne.
AREA: 1 mile by ½ mile.
POPULATION: uninhabited.
ACCESS: weather permitting, by hired boat from Luing, Easdale, or Mull.

Sometimes called Hinba Island, Eileach An Naoimh is the most southerly of the Garvellochs group (q.v.), and it is famous because three great seafaring monks, St Brendan of Clonfert, St Columcille of Iona, and Cormac, were once all at the tiny retreat there. According to Adamnan, St Columba is also said to have made several retreats to the island.

The shores are chiefly low cliffs or steep rocks, and a dinghy landing should be made at Port Choliumchille, a rocky inlet in the SE. cliffs. You will need to pull your craft well up on to the shingly beach.

Eileach An Naoimh is unique in that it possesses examples of some of the earliest of all Celtic-Christian remains, notably the beehive cells of the early *papae*, older than any on Iona, the ruins of a tiny chapel, and a graveyard. These form the last surviving architectural link with the Celtic monastic development in the W. of Scotland. The remains of the graveyard suggest that priests were buried here, but more interesting is the small circle of stones, on one of which is a rough cross, a short distance to the S. of this. Legend has it that this was the burial place of St Columba's mother.

The well at the head of the landing place still contains cold, drinkable water.

EILEACH CHONAIL

SITUATION: the most northerly isle of the Garvellochs group, close to the island of Luing and 5 miles W. of the Argyll coast at the entrance to the Firth of Lorne.
AREA: 1¼ miles by ½ mile.
POPULATION: uninhabited.
ACCESS: by boat from Luing or Easdale.

This isle takes its name from Dun Chonail Castle, built there in the 13th century, the ruins of which still remain. It was at one time inhabited by a fair-sized colony.

See also Garvellochs, the.

EILEAN A'GHOBHA: *see under* Flannan Isles.

EILEAN AN TIGHE

SITUATION: the largest of the Shiant Isles, linked by a narrow spit of stony beach to the island of Garbh Eilean.
AREA: 350 acres.
POPULATION: uninhabited.
ACCESS: by boat from Lewis (6 miles) or Skye (14 miles).

Though frequently regarded as two islands, Eilean An Tighe and Garbh Eilean are really one. The best way of making a landing here is by rowing boat from a boat anchored close inshore to the spit which joins the two parts. It is easy to pull the rowing boat up on to the pebble beach.

There are tall cliffs rising to 400 ft in the NE., giving some shelter to the Bay of Shiant. Near the landing place is a cottage once owned by Sir Compton Mackenzie, a well-built Hebridean stone house with whitewashed walls. There is a well with a spring not far away. There are ruins of other cottages elsewhere on the island, which a century ago was inhabited. There is also the site of an ancient church, but little remains to distinguish it.

The extensive bird life is described in this book under the Shiant Isles (q.v.), but what has not been mentioned there is the presence of the eider duck, which is recorded as having been seen on Eilean An Tighe by L. R. Higgins.

The N. of the island is more hilly than the S., rising in one place to more than 500 ft. In the lower-lying portions of the island water mint, buttercups, meadowsweet, forget-me-not and willow withy grow. There are also marsh marigolds, orchis, and masses of Scotch thistle.

EILEAN BAN

SITUATION: in the narrow straits between Kyleakin on the SE. coast of Skye and the Kyle of Lochalsh.
AREA: 12 acres.
POPULATION: uninhabited.
ACCESS: by boat from Kyleakin.

Few tiny islands can be more fascinating than this one, sometimes called 'The White Island' on account of its dazzling white sands. John Buchan in *A Prince of the Captivity* wrote of it:

It was different from the island that he remembered. There were the white sands that he knew, and the white quartz boulders tumbled amid the heather. . . . But the place seemed to have grown larger. . . . It seemed to him that the thymy downs now extended for ever. He had stridden over them for hours and had found delectable things – a new lochan with trout rising among yellow water-lilies, a glen full of alders and singing waters, a hollow with old gnarled firs in it and the ruin of a cottage pink with foxgloves. But he had never come within sight of the sea, though it seemed as if the rumour of its tides was always in his ear.

If it charmed John Buchan, it also captured the imagination of the author Gavin Maxwell, who bought it and aimed to convert it into a zoo for birds and mammals of the West Highlands. With Mr John Lister-Kaye he brought this project into being, stocking the island with herons, gannets, carrion crows, owls, a ram and two goats, and other animals. His dream of an even more extensive zoo was brought to an end by his untimely death. There are no signs that the island is being developed in any way at present, but it still holds the Commissioners of Northern Lights' automatic light.

Book: *The White Island*, by John Lister-Kaye, Longmans, 1972.

EILEAN BEAG: *see under* Crowlin Isles.

EILEAN CHATHASTAIL

SITUATION: 200 yds S. of Eigg in the Inner Hebrides and 12 miles W. of Arisaig on the Scottish mainland.
AREA: 8 acres.
POPULATION: uninhabited.
ACCESS: by boat from Galmisdale, Eigg.

The island is of interest to those visiting Eigg (q.v.) because it has on it the remains of an ancient mausoleum.

EILEAN CHOINAID

SITUATION: the smallest of the Summer Isles, off the coast of Ross and Cromarty.
AREA: 6 acres.
POPULATION: uninhabited.
ACCESS: by boat from Achiltibuie.

EILEAN GARBH: *see under* Badcall.

EILEAN MEADHONACH: *see under* Crowlin Isles.

EILEAN MHUIRE

SITUATION: the smallest of the Shiant Isles, 6 miles SE. of Lewis in the Outer Hebrides, and 14 miles N. of Skye.
AREA: 30 acres.
POPULATION: uninhabited.
ACCESS: by boat from Lewis or Skye.

Eilean Mhuire means 'Mary's Island', and the island is the site of an ancient church dedicated to the Virgin Mary. It is separated by a strait from the spit-linked islands of Eilean An Tighe (q.v.) and Garbh Eilean.

See also Shiant Isles.

EILEAN MOR (Crowlin Isles): *see under* Crowlin Isles.

EILEAN MOR (Flannan Isles)

SITUATION: the largest of the Flannan Isles, about 22 miles off the west coast of Lewis in the Outer Hebrides.
AREA: 39 acres.
POPULATION: uninhabited except for lighthouse keepers.
ACCESS: by boat from Uig Bay, Lewis.

Eilean Mor is best known for its lighthouse, called the Flannans Lighthouse, built between 1895 and 1899. This is one of the remotest lighthouses in the British Isles, if not in the whole world. It has a staff of four men, of whom one acts as relief keeper and spends his off-duty periods of three weeks at the shore station on Lewis. The lighthouse, which has a group-flashing white light, giving two flashes in quick succession every thirty seconds, is 75 ft high and stands some 330 ft above sea-level.

The lighthouse was the scene of one of the strangest sea mysteries of modern times. On 15 December 1900 a steamer, the *Archer*, sailing from Philadelphia to Leith, passed Eilean Mor and reported that there was no beam from the lighthouse. The Commissioners of Northern Lights' relief ship set out to investigate, but owing to strong gales did not reach the Flannans until Boxing Day. When the relief keeper went ashore he found no trace of the other three keepers. All was neat and tidy and the lamp was in order, but cold and unlit. Entries in the log had been made up to 13 December, but they gave no clue as to what had happened. A search of the island was made without success. The mystery remains unsolved to this day.

There are two landing places on Eilean Mor built up from concrete blocks cemented against the cliffs. A crane above the main landing place takes stores up from the boat. Situated close to the lighthouse is a ruined chapel dedicated to St Flannan, built of dry stone and called the Blessing House. The walls are about $2\frac{1}{2}$ ft thick, but the building is very small and primitive, with a low doorway at the west end. There are also some other primitive buildings which at one time must have served as accommodation for the Lewis crofters when they visited the island.

Eilean Mor rises to a height of 288 ft and is easily the highest point in all the Flannan Isles. There is a curious tradition of sanctity about this island. In the past the legend of Eilean Mor's 'holiness' had a restraining influence on the conduct of the men of Lewis when they visited the island. They observed certain strict rules which did not apply on Lewis, and made a point of going to the ruined chapel to pray before they went fowling.

See also Flannan Isles.

EILEAN MOR (Loch Sween)

SITUATION: in the mouth of Loch Sween, an arm of the sea extending from Kilmichael to Kilmory, Argyll.
AREA: 20 acres.
ACCESS: by boat from the Scottish mainland.

On Eilean Mor, one of many tiny islands bearing this name off the W. coast of Scotland, are the ruins of an ancient chapel and oratory of St Cormac. The tombstone of a priest wearing robes and some grotesque figures may be seen.

EILEAN NA BEARACHD: *see under* Badcall.

EILEAN NA NAOIMH

SITUATION: in the Kyle of Tongue, Sutherland, 1 mile offshore.
AREA: 30 acres.
POPULATION: uninhabited.
ACCESS: by boat from the Scottish mainland.

Eilean Na Naoimh means 'Saint's Island', and the island formerly had a chapel and burial place on it, traces of which are still to be seen. On the S. side of the island the sea, after passing for several yards through a narrow channel, spouts up into the air, sometimes to a height of 30 ft, through a hole in the rock.

EILEAN NA RAINICH: *see under* Badcall.

EILEAN NA SAILLE

SITUATION: one of the Summer Isles, off the Ross and Cromarty coast, 100 yds from the NW. coast of the island of Tanera More.
AREA: 12 acres.
POPULATION: uninhabited.
ACCESS: at low tide on foot across the mud flats from Tanera More.

A tiny island, but nevertheless an attractive one with low cliffs, never more than about 20 ft, rising to grass terraces, and a tiny hillock of bracken in the centre. There is not much bird life except for a few oyster-catchers and occasionally ringed plovers or Arctic terns.

See also Summer Isles.

EILEAN NAN EACH

SITUATION: $\frac{1}{4}$ mile NW. of Muck in the Inner Hebrides.
AREA: 75 acres.
POPULATION: uninhabited.
ACCESS: by boat from Muck.

Sometimes called Horse Island, Eilean Nan Each is a grassy isle, rising to 178 ft. It has good grazing land and is rich in bird life, including herons. Wild orchids can sometimes be found here.

EILEAN TIGHE: *see under* Flannan Isles.

EILEANS (Bute), the: *see under* Cumbraes, the.

ENSAY: *see under* Sound of Harris Islands.

EORSA

SITUATION: in Loch Na Keal, $\frac{1}{4}$ mile off the W. coast of Mull in the Inner Hebrides.
AREA: 1 mile by $\frac{1}{2}$ mile.
POPULATION: uninhabited.
ACCESS: by boat from Torloisk, but no regular service.

Eorsa once belonged to the Prioress of Iona. Though uninhabited, it has a stock of sheep, and was formerly the domain of wild goats. It is rugged and landings are not always possible.

ERISGEIR

SITUATION: W. of the Ardmeanach promontory, Isle of Mull, in the Inner Hebrides.
AREA: 3 acres.
POPULATION: uninhabited.
ACCESS: by boat from Ardtun.

This islet is little more than a flat-topped rock on which sheep can graze. Of all the islands around Mull it is the smallest and least significant. Yet it was destined to play a part in the history of Mull by the very fact of its insignificance. In the 14th century the great power in the Hebrides was the Lord of the Isles, whose possessions included all the islands around Mull. His daughter was married to the chief of the Clan Maclean, who at that time had few possessions himself. When their son was born the Lord of the Isles said he wished to make a gift of land to his grandchild. The wily father, sensing that the Lord of the Isles had in mind some trivial plot, cunningly suggested that an admirable gift would be that of 'little Erisgeir and her isles'. The Lord of the Isles agreed with alacrity, knowing full well that Erisgeir was a tiny island, and a contract was drawn up. It was only afterwards that he realised that the phrase 'her isles' included the large and fertile island of Mull itself.

Thus were founded the fortunes of the Clan Maclean.

ERISKAY

SITUATION: in the Sound of Barra between Barra and South Uist in the Outer Hebrides. It is ¾ mile S. of South Uist and 5 miles from Barra.

AREA: 3 miles by $1\frac{1}{2}$ miles.
POPULATION: 280.
ACCESS: by passenger ferry service or by locally hired boat. Communications with Barra are infrequent, but there is a fairly regular service from Pollachar Inn, South Uist when weather permits.

This is the island on which 'Bonnie Prince Charlie' landed when he first set foot in Scotland in 1744 prior to his rebellion of the following year. The French frigate, *La Doutelle*, after surviving many fierce storms finally sighted the Outer Hebrides, and it was on the Prince's order that a landing was made on Eriskay. He was brought ashore in a long-boat and the strand on which he landed has been known ever since as Coilleag d'Phrionnsa, or 'the Prince's Strand'. It is said that the pink convolvulus that flowers in the bay grew from seeds which he scattered as he landed.

History records that the Prince did not receive much encouragement for his enterprise from the inhabitants of Eriskay, though he was given a warm if primitive welcome. But ever since the legends of the 'Bonnie Prince' have been maintained in the island. The pink convolvulus is called 'The Prince's flower' and the Eriskans insist that it will not grow anywhere but on Eriskay. This may not be absolutely true, but Alasdair Alpin MacGregor records that 'Friends of my own have transplanted roots of it from Eriskay to their mainland gardens, but without success. Such transplantings have usually withered and died before the ensuing spring, thus confirming – and to their satisfaction – what the Eriskay folk foretold.'

But perhaps Eriskay's proudest boast is that it was on this island that drambuie, the celebrated drink of the 1745 Rebellion, was first drunk on Scottish soil. The secret recipe of drambuie came with the Prince from France and he himself introduced it to Scotland, leaving his secret behind him. But the story of what happened to that secret recipe and how it was developed belongs properly to Skye, and is duly recorded in the section of this book dealing with that island.

Eriskay, like other neighbouring islands, has held on to its population while people in most other islands in the Hebrides have drifted to the mainland. All its people are Roman Catholics, which probably explains why the Jacobite tradition is so strongly maintained. For many years the island was known as 'Father Allan's Isle' after the parish priest, Father Allan MacDonald who, with the islanders' co-operation, built Eriskay's only church, which was opened in 1902. He was the inspiration of his flock and is remembered and spoken of to this day, not merely for his spiritual guidance but for what he did to encourage the perpetuation of Gaelic folklore. He even published a book entitled *Gaelic Words and Expressions from South Uist and Eriskay*.

His influence lives on, for Eriskay was made famous (justly) for its folk-songs through the efforts of her priest, and is probably the one island in the Hebrides where the genuine folk-songs can still be heard, sung by the women as they knit the fishermen's jerseys. Indeed the folk-lore even extends to the knitting, for the patterns on each jersey symbolises the Tree of Life. Many of the ancient songs have been carefully written down and preserved, including many love-songs, herd-songs, and rowing-songs – not forgetting the celebrated 'Eriskay Love Lilt', a Gaelic melody that is known far beyond the confines of Scotland.

It is an island of much beauty, of soft and varied colours, and gleaming white sands which are often inhabited by sandpipers and plovers. There are no hotels on Eriskay, but many of the crofters take in visitors in summer. Most of the inhabitants fish for a living, for lobsters especially. But Eriskay has managed to keep aloof from the tourist, for whom there is no form of entertainment, no angling, and no golf course. There remains on the island a good strain of the famous Eriskay pony, and colonies of seals can often be seen crossing from South Uist. Some magnificent shells are to be found on the beaches and Coilleag, the tiny hamlet above the Prince's Strand, possesses some attractive old thatched cottages.

See also Calvay.

Accommodation: there are some cottages to let on Eriskay in summer. There is no inn.

Post Office.

Books: *The Enchanted Isles*, by Alasdair Alpin MacGregor, Michael Joseph, 1967; *Father Allan's Island*, by Amy Murray, Moray Press, 1936.

ERRAID

SITUATION: ¼ mile W. of SW. end of the Ross of Mull in the Inner Hebrides.

AREA: 1 sq. mile.

POPULATION: 29.

ACCESS: at low tide it is possible to walk across to Erraid from Mull.

Erraid was immortalised by Robert Louis Stevenson in *Kidnapped*. After being kidnapped and taken aboard the brig *Covenant*, the youthful David Balfour tells how he is washed overboard

off a little isle they called Erraid, which lay low and black upon the larboard. . . . The whole, not only of Erraid, but of the neighbouring part of Mull (which they call the Ross) is nothing but a jumble of granite rocks with heather in among. . . . At last I came to a rising ground, and it burst upon me all in a moment that I was cast upon a little barren isle, and cut off on every side by the salt seas. . . . The second day I crossed the island to all sides. There was no one part of it better than another; it was all desolate and rocky; nothing living on it but game birds which I lacked the means to kill, and the gulls which haunted the outlying rocks in a prodigious number. But the creek, or straits, that cut off the isle from the main land of the Ross, opened out on the north into a bay and the bay again opened into the Sound of Iona; and it was the neighbourhood of this place that I chose to be my home; though if I had thought upon the very name of home in such a spot, I must have burst out weeping.

I had good reasons for my choice. There was in this part of the isle a little hut of a house like a pig's hut, where fishers used to sleep when they came there upon their business; but the turf roof of it had fallen entirely in; so that the hut was of no use to me, and gave me less shelter than my rocks.

Stevenson must have discovered this island as a young man when his father, Thomas Stevenson, engineer to the Commissioners of Northern Lights, came there to supervise the construction of the two lighthouses of Skerryvore and Dhu Heartach (q.v.), situated several miles distant on dangerous reefs. For Stevenson senior used Erraid as a base for his operations. Indeed, until quite recently Erraid was the shore station for the keepers of these lighthouses, who lived in houses on the island. This property, including nine cottages, was sold in 1967 by the Commissioner of Northern Lights.

Stevenson's David Balfour speaks with feeling about 'that miserable isle' and how he did not at first realise that at low tide he could quite easily cross to the main island of Mull: 'A sea-bred boy would not have stayed a day on Erraid; which is only what they call a tidal islet, and, except in the bottom of the neaps, can be entered and left twice in every twenty-four hours, either dry-shod, or at the most by wading.'

Erraid is sometimes spelt Earraid and Erraidh. It really is as bleak as Stevenson describes it, and lies in a particularly dangerous area for shipping, with the Torran Rocks only 3 miles to the S. At the time of going to press Erraid is up for sale 'with vacant possession', and it is stated that it possesses 'a small house, 8 cottages, sheep farm and first-class lobster fishing'.

EWE

SITUATION: in the middle of Loch Ewe, Ross and Cromarty.
AREA: 2¼ miles by 1 mile.
POPULATION: 5 (1970).
ACCESS: by boat from the Scottish mainland.

Ewe Island, or Isle of Ewe, as it is known locally, is low-lying and mainly covered with grass and heath, though Dean Munro visited it and declared that in 1549 it was 'pleasantly wooded'. Its highest point is 233 ft.

Sir Francis Alexander McKenzie of Gairloch paid much attention to developing this island in the last century. Today it is used for farming, both arable and hill grazing, and the owners use it as a base for lobster and prawn fishing. Two families were reported to be living there recently. There are two houses and a small jetty.

Permission to visit the island should be obtained from the owners, Grant Bros, Isle Ewe, Aultbea, near Ross.

EYNHALLOW

SITUATION: in Eynhallow Sound, midway between Rousay and Orkney Mainland.
AREA: 1 mile by ¾ mile.
POPULATION: uninhabited.
ACCESS: by boat from Rousay or Orkney Mainland.

A tiny island worthy of mention merely because it contains the ruins of the church of a monastery that once gave an abbot to Melrose.

John Mooney carried out intensive research over a lengthy period into the history of Eynhallow (which means literally 'Holy Isle'). The ruins of the former church are barely discernible, and the island is only rarely visited.

See also Orkney Islands.

Book: *Eynhallow*, by John Mooney, Orcadian, Kirkwall, 1923.

FAIR ISLE

SITUATION: midway between the Orkneys and the Shetlands, 30 miles NNE. of North Ronaldsay and 28 miles SSW. of Sumburgh Head in the Shetlands.

AREA: 3 miles by 1½ miles.

POPULATION: 80 (1972).

ACCESS: by air to Sumburgh, Shetland, from Aberdeen (BEA), or from Edinburgh, with connections from the S.; or by sea to Lerwick, in Shetland, from Aberdeen or Leith. The boat to Fair Isle leaves from Grutness, Shetland. If you fly it is necessary to spend the night at one of the Grutness hotels. From May to September the Fair Isle boat, *Good Shepherd*, leaves Grutness at 11.30 am on Tuesdays and Fridays. Loganair runs charter flights from Kirkwall, Orkney, to Fair Isle.

This is the most isolated inhabited island in the British Isles, and was acquired in 1954 by the National Trust for Scotland thanks to a grant of £5,500 from the Dulverton Trust. Until that date it seemed inevitable that the island must lose its remaining population, but the National Trust, with the co-operation of the Zetland (Shetland) County Council and the Crofters' Commission, saved the situation by building a new jetty, modernising houses, and providing electricity. The development of the Bird Observatory, now the foremost in Europe, has also attracted an increasing number of visitors.

The name Fair Isle has various possible Norse connotations – *Faerey*, 'the Sheep Isle', *Feoer*, meaning 'far off', and *Fara*, 'the far isle'. It was obviously one of the

Fair Isle (J. Allan Cash)

'beacon islands' during the Vikings' occupation, and in the 12th century was farmed by one Dagfinn. In 1588 *El Gran Griffon*, a ship of the Spanish Armada, was wrecked on Fair Isle and the islanders had 300 men to feed in addition to themselves. During their stay on the island the Spaniards taught the inhabitants how to make fast dyes from flowers, a secret which has been kept to this day. A visiting minister in 1700 reported that there were only some ten families living on Fair Isle, as 'two-thirds of the population have been swept away by smallpox'. Yet in 1861 the population was 380, though the people were living in conditions of dire poverty, their diet being coarse home-ground meal and fish. They lived in earth huts.

Eventually authorities on the mainland concerned themselves with the plight of the islanders, and in 1862 some 137 people left Fair Isle for Canada. Brutal landlordism exacerbated the conditions of the people in the 19th century, but it is worth noticing that when, in 1876, a German ship was waterlogged off Fair Isle and the islanders performed heroic feats in rescuing the personnel, the German Government presented the helmsman of the rescuing boat with a watch, and his crew with money. Until there was an organised mail service to the island the inhabitants sailed to the Shetlands in open boats with boxes of knitting to sell. They went to Orkney to barter and buy seed and meal, and to Westray in the Orkneys to be married.

The Bird Observatory is at North Haven in the NE. of the island and the main crofting area is in the S., around South Harbour. Fair Isle itself is a combination of green fields, moors, and sandstone cliffs. It has a school, a shop, and two churches, one Church of Scotland and the other a Methodist Chapel. Alongside the latter is a hall where films are shown and dances held. Across the fields from the hall is Kirkigeo, where the village graveyard is situated.

The main occupation is crofting, chiefly sheep rearing, and other islanders obtain part-time employment manning the mail-boats, fishing, and doing coastguard work. The women contribute largely to Fair Isle's livelihood by their weaving and knitting of the world-famous Fair Isle woollen garments. Their colourful patterns have acquired an international reputation. Ten years ago more than fifty families from all over Britain applied for two crofts on Fair Isle when the National Trust invited applicants in an effort to boost the island's dwindling population.

There is nothing of special archaeological interest on the island except possibly for the remains of an old watermill at Finniquoy, where the islanders used to grind their meal, and a watch-tower dating from Napoleonic times on Malcolm's Head.

The Bird Observatory was first established in 1948 when an ex-naval station at North Haven was converted for this purpose. Since then more than 300 species of birds have been recorded here, and about 62,000 birds have been ringed by the Observatory staff. Some of those ringed have been found as far afield as the Greek Islands (a rustic bunting), Siberia (a bar-tailed godwit), Brazil (an Arctic skua), and the Canadian Arctic (a snow bunting).

Thousands of sea-birds arrive on Fair Isle in the summer, though most of the rare species are only to be seen in spring and autumn. Puffins predominate, but there are also fulmars, black guillemots, razorbills, kittiwakes, and herring gulls in large numbers. In 1968 four young eagles were imported to Fair Isle from Norway, and of these three have survived. On the moors may be found the bonxie and the Arctic skua, and on the W. cliffs the visitor may come across the peregrine falcon – a few pairs nest there each year. Other nesting birds include eiders, oyster-catchers, ringed plovers, wheatears, rock and meadow pipits, ravens, hooded crows, corncrakes, and the Fair Isle wren.

Grey seals breed in some of the sheltered coves and move on to the skerries near the North Light in summer. Rabbits abound – black, white, and brown – and the island also has its own Fair Isle field-mouse. There is nothing distinctive about the plant life, but in summer the island is covered with blue squills, sea pinks along the cliff faces, red and white campion, and willows.

See also Shetland Isles.

Accommodation: the Fair Isle Bird Observatory offers double and single rooms with hot and cold water, a lounge and a library, with full board – details from The Warden, Bird Observatory, Fair Isle. Guidance and instruction on natural history from the staff is provided. Rooms are also let by some of the islanders. There are no licensed premises on the island.

Post Office.

Book: *Fair Isle*, published by W. S. Wilson, Stackhoull Stores, Fair Isle.

FARA (Hoy)

SITUATION: in Scapa Flow in the Orkney Islands, 2 miles E. of Hoy.
AREA: 200 acres.
POPULATION: 4.
ACCESS: by boat from Hoy or Kirkwall.

Before World War II there were twenty-five people living on Fara; today three of the population of four are women. They all live off their small farm and lobster fishing.

Mr Gordon Watters, the present owner, says that the main reason for the decline in population is that six of the men previously on the island never married: 'bachelors have been the ruin of Fara.' He has three fields under cultivation for oats, turnips (for fodder), cabbages, and potatoes, and seaweed is used for manure. The island is bleak and grass for grazing is scarce, much of the land being covered with heather. There are, however, some sheep.

Cuddy is also fished in the vicinity, salted, dried in the wind, and smoked over a peat fire. The islanders eat the cuddy themselves.

Mrs Watters is Fara's postmistress and is in charge of the one telephone box on the island, half a mile from her home. This is the sole link with the outside world.

There are many gulls' nests on the headland at the N. end of Fara.

See also Orkney Islands.

FARA (Westray)

SITUATION: 3 miles SE. of Westray in the Orkney Islands.
AREA: 180 acres.
POPULATION: uninhabited.
ACCESS: by boat from Westray.

The island is used for grazing. *See also* Orkney Islands.

FETLAR

SITUATION: 2½ miles E. of Yell and 3 miles S. of Unst in the Shetlands.
AREA: 14 sq. miles.
POPULATION: 224.
ACCESS: by boat from Yell or Unst.

This is the most fertile of all the Shetland Islands (q.v.), noted for its ponies and containing much of archaeological interest. Its main centre of population is at the Wick of Tresta in the S. of the island. There are high cliffs around much of its coastline, and the Wick of Gruting in the N. sometimes provides shelter for Russian fishing vessels during the winter storms.

In the NW. corner of Fetlar, on Hamara Field, is a semicircular enclosure of large serpentine stones, about 3–4 ft high: in the centre of this are three earth-fast stones. Between Skutes Water and Vord Hill are to be found the Fiddler's Crus: three rings of small stones set edge to edge at the corners of a triangle. The *Inventory of Monuments* states that 'no satisfactory suggestion as to the period or purpose of these curious constructions can be offered.' Not far away is an outer ring of serpentine stones, about 36 ft across, and inside this a low earth-bank with two earth-fast stones in the centre.

Study of these and other antiquities of Fetlar should be rewarding for any archaeologist – not least the remains of a stone dike that once divided the island into two halves. Its name is Funzie Girt, meaning 'the Finns' Dike'. There are also some *brochs*, mounds, and cairns, all proof of the fact that the fertile pasture land of Fetlar attracted settlers in ancient times.

Fetlar's chief disadvantage is that it lacks a good harbour, but despite this it is well worth a visit, and the ponies alone are a considerable attraction.

A short while ago Mr Robert Stenuit went to Fetlar to try to retrieve the cargo of an 18th-century Danish frigate, the *Wendela*. This ship, wrecked off Fetlar in 1737, was laden with several tons of gold and silver in ingots and coins. Mr Stenuit examined all the documents concerning the ship and its cargo and all previous reports of attempts to salvage it. Eventually he recovered a quantity of silver pieces, including pieces of eight, a few ducatoons and patagoons from Holland, and some Scandinavian coins.

Book: *Orkney and Shetland*, by Eric Linklater, Robert Hale, 1965.

FIDRA

SITUATION: 2½ miles WNW. of North Berwick in East Lothian.
AREA: 15 acres.
POPULATION: uninhabited.
ACCESS: by special arrangement from North Berwick.

A tiny island which holds the remains of a chapel dedicated to St Nicholas, dating back to the 13th century. There is not much else of interest, except that it is an island capable of limited agricultural development.

FLADDA

SITUATION: one of the Treshnish Isles, 3½ miles SW. of Mull in the Inner Hebrides.
AREA: 90 acres.
POPULATION: uninhabited.
ACCESS: by boat from Mull.

The second largest of the Treshnish Isles (q.v.), Fladda is almost divided into two islands by a narrow causeway. It is flat-topped, with cliffs rising to 70 ft, and in its cliff area has a rock pavement on which the seals come to breed in autumn.

It has fresh water, like nearby Lunga (q.v.), and many rock pools. Though there are some wild flowers here, the bird life is less varied than on Lunga. Shags, a herring gull colony, some wheatears and pipits: that is about the sum total of bird life.

The best landing place lies at a beach situated at the head of a long creek, not easy to reach at low tide.

FLADDAY

SITUATION: close inshore to Raasay Island, which lies to its E., and 4 miles E. of Skye.
AREA: 2 miles by 1 mile.
POPULATION: uninhabited.
ACCESS: by boat from Raasay.

The island is used for grazing.

FLANNAN ISLES

SITUATION: about 22 miles off the west coast of Lewis in the Outer Hebrides.
TOTAL AREA: 97 acres.
POPULATION: 4.
ACCESS: by boat from Uig Bay, Lewis.

Sometimes known as the Seven Hunters, the seven islands of Flannan are Eilean Mor (q.v.), Eilean Tighe (18 acres), Eilean a'Ghobha (12 acres), Soray (8 acres), Roareim (7 acres), Sgeir Toman (5 acres), and Sgeir Righinn (3 acres). There are also several smaller islets and rocks which total only about 3–4 acres between them.

The Flannan Isles belong to the parish of Lewis. The largest of them have grassy surfaces which make excellent grazing ground for sheep, and they are used for this purpose by Lewis crofters. There is a picturesque description of the islands by Dr J. MacCulloch in his *Description of the Western Isles* in 1824: 'like a meadow thickly enamelled with daisies'. All the Flannans are rich in bird life, more than a hundred species having been recorded there, including Leach's fork-tailed petrel, the fulmar, shag, and kittiwake. Plant life is scarce.

Landings are exceedingly difficult on all islands except Eilean Mor, and are possible only in favourable weather conditions. The rock-strewn sea around the Flannans makes navigation hazardous. The islands are reputed to have taken their name from one of two saints – Flannan, Bishop of Cell da Lua in Eire, or St Flann, son of an Abbot of Iona who died in 891. John Morison, writing in the 17th century, stated that their name then was the Isles of 'Sant Flannan . . . where Sant Flannan himself lived ane hermit.'

See also Eilean Mor (Flannan Isles).

Books: *Description of the Western Isles*, by J. MacCulloch, 1824; 'Natural History Notes from Certain Scottish Islands', an article by R. Atkinson in the *Scottish Naturalist*, 1938.

FLEET, ISLES OF: *see* Isles of Fleet.

FLODDAY: *see under* Barra Isles.

FLOTTA

SITUATION: in Scapa Flow in the Orkneys, 2½ miles E. of Hoy.
AREA: 3 miles by 2 miles.
POPULATION: 130 (1970).
ACCESS: by boat from Lyness, Hoy.

Flotta, as its name suggests, is a low-lying island and is strategically placed at the S. entrance to Scapa Flow. For this reason it was turned into a site for barrage balloons and anti-aircraft guns in World War II, an operation which swiftly changed an old-world farming estate, on which oxen still drew the plough, into a miniature garrison. When the war was over Flotta had roads and two concrete piers. Today there are shops, a primary school, and a resident doctor on the island.

The farms here are small, but the community is contented and ekes out a living with lobster-fishing as well as agriculture. In 1970 at a cost of £13,000 a submarine pipeline was laid to bring water to Flotta from Hoy.

See also Orkney Islands.

FOULA

SITUATION: 16 miles W. of Shetland Mainland and 35 miles N. of the Orkney Islands.
AREA: 3 miles by 2 miles.
POPULATION: 50 (1970).
ACCESS: by sea from Walls, Shetland.

The remotest of the Shetland Islands (q.v.), Foula has magnificent cliffs rising to 1,220 ft, and can only be reached in relatively good weather, owing to the smallness of the island's only harbour, Ham Voe.

A sense of total isolation has tended to reduce the population steadily over recent years: in the last century there were three times as many inhabitants. The cliffs rise to five peaks, the highest of which is the Sneug (1,373 ft). The Kame, the second highest sea-cliff in the British Isles, is situated in the NW. Around Ham Voe is the island's main settlement, which constitutes a shop, a post office, a few houses, and a school. There is another tiny settlement at Hametoon, near South Ness.

The Norse language was in use on the island until the beginning of the 19th century.

By far the most interesting feature of Foula is its large colony of skua gulls, or bonxies, which breed here. These aggressive birds indulge in dive-bombing attacks on human beings during the breeding season, and can be extremely vicious. This is an ideal place in which to observe the habits of this rare species. Kittiwakes, red-throated divers, and all kinds of gulls are also to be found on Foula. It was here that the film *Edge of the World* was made.

FUDAY

SITUATION: off Northbay, Barra, in the south of the Outer Hebrides.
AREA: 5 acres.
POPULATION: uninhabited.
ACCESS: by boat from Northbay, Barra.

Of some interest to archaeologists on account of the Dunan Ruadh ancient remains and burial chambers, Fuday is a mixture of sand dunes and remarkably good pasture land. At one time this tiny island is reputed to have supported some eighty cattle all the year round. It is also remarkable for the variety of its wild flowers, and the barnacle geese which usually arrive on the island in mid-October, remaining until April or May. Then, on a signal from their leaders, they rise together in one tremendous flock and set off towards the Arctic in wedge-shaped formations.

Late in the last century a great gale struck Fuday and, in sweeping the sand dunes, exposed to view five stone cells with the skeleton of a man in each of them. There is a well on Fuday known locally as Tobar nan Ceann, which means 'The Well of Heads'. It was into this well, according to legend, that the Macneil of Barra hurled the heads of the vanquished Norsemen after defeating them in battle on the island.

See also Barra Isles.

FUIARY

SITUATION: off Northbay, Barra, in the south of the Outer Hebrides.
AREA: 7 acres.
POPULATION: uninhabited.
ACCESS: by boat from Barra.

Fuiary is very close to Fuday and, like that island, is visited by the barnacle geese. Apart from that its only interest is for students of the occult, on account of the fact that it is said to be inhabited by a fairy washerwoman, whom some of the inhabitants of Barra claim to have seen in quite recent years. John Macpherson, one of the most famous of Barra's recent inhabitants, claimed that on one of her appearances the fairy washerwoman was seen descending from the highest part of the island to the burn, where she proceeded to wash a death-shroud.

See also Barra Isles.

GAIRSAY

SITUATION: 1 mile E. of Orkney Mainland.
AREA: 2 miles by $1\frac{1}{2}$ miles.
POPULATION: 28.
ACCESS: by boat from Finstown.

An oval, brown mass rising from the sea, Gairsay provides some shelter to Wide Firth between Orkney Mainland (q.v.) and Shapinsay (q.v.). It was formerly the abode of an adventurous Norseman, Sweyn Asleifsson, who combined raiding with harvesting. The *Orkneyingers' Saga* states that 'in the spring' Sweyn

> worked hard and made them lay down very much seed, and looked after it himself. But when that toil was ended, he fared away every spring on a viking voyage and harried much about the Southern Isles and Ireland and came home after midsummer. That he called his spring-viking. Then he was home till the corn fields were reaped and the grain seen to and stored. Then he fared away on a viking voyage and did not come home again till winter was one month spent, and that he called his autumn-viking.

This 12th-century Norse chieftain built himself a drinking-hall on the island where he lived with some eighty retainers. No trace of this now exists.

In 1963 the island was advertised as being for sale.

See also Orkney Islands.

GARBH EILEAN: *see under* Eilean An Tighe.

THE GARVELLOCHS

SITUATION: at the approach to the Firth of Lorne, 5 miles W. of the Argyll coast.
AREA OF ISLANDS: 3 miles by not more than ½ mile at the widest point.
POPULATION: uninhabited.
ACCESS: by hired boat from the island of Luing, or Easdale.

The Garvellochs are a chain of small islands, sometimes called 'The Isles of the Sea' and quite often 'The Rough Isles'. They are aptly named, for the sea is apt to be rough in these parts and a landing can only be made on the islands in good weather conditions.

Their chief significance is their visible links with ancient history: for so small a group of islands they are rich in archaeological interest. Certainly they possess evidence of occupation dating back to the earliest Christian period, far earlier than that of Iona.

See also Eileach An Naoimh, Eileach Chonail.

GIGALUM

SITUATION: ½ mile off the SE. tip of Gigha Island, Kintyre.
AREA: below 20 acres.
POPULATION: uninhabited.
ACCESS: from Gigha by special arrangement.

The tiny island of Gigalum belongs to the owners of Gigha (q.v.), the family of the late Lieutenant-Colonel Sir James Horlick. It is suitable for grazing, and was at one time used by crofters.

GIGHA

SITUATION: 3 miles W. of Kintyre on the Scottish mainland.
AREA: 9 sq. miles.
POPULATION: 180 (1972).
ACCESS: daily boats are run from Kennacraig to Gigha by Western Ferries. A steamer service calls at the island from Tarbert on Tuesdays, Thursdays, and Saturdays, returning on Mondays, Wednesdays, and Fridays; for passengers only there is a small ferryboat service several times a day from Tayinloan.

The Isle of Gigha is full of fascinating contrasts, ranging from the gaunt, rocky bays on its W. coast to the soft, lush farmland and wooded gardens on the E. and the long line of very low hills which runs from N. to S. for about 6 miles. There are splendid views from the hilltops to Mull, Jura, Kintyre, and Arran. The largest hill on Gigha, Creag Bhan (331 ft), makes an admirable vantage point for the whole island.

An Ogam stone on the island, dating from the 5th century, proves occupation at that time by the Celts. Gigha was under Norse rule for about 400 years, and Hakon of Norway landed here in 1263 during his disputes on the ownership of Scotland with Alexander III. The ownership of the island alternated between the Macdonalds and the MacNeills of Kintyre and Colonsay from about 1344 until the 19th century. In 1944 Gigha was bought by the late Lieutenant-Colonel Sir James Horlick, who transformed it both in the economic and environmental sense. Gigha today is a perfect example of what a single landlord can achieve for the benefit of all.

Dairy farming on the island has been modernised so that a milk production of 70,000 gallons a year in 1944 has risen to 300,000 gallons today. The Gigha Creamery Company was launched with the co-operation of tenant-farmers, and a cheese-making factory established. Sir James Horlick has had three 10,000-gallon fuel tanks

Gigha (Aerofilms)

installed at the pier so that the islanders can buy their supplies in bulk at low cost, a facility which has also been made available to fishermen who call in regularly at the island port.

There is one hamlet on the island, Ardminish, with an island store and a Church of Scotland church with resident minister. A small, richly-coloured stained-glass window in the S. wall of Gigha's parish church is a memorial to one of the island's best-known former residents, Kenneth MacLeod, one-time minister here, but better known as a poet, a collector of old Gaelic songs and tunes, and the composer/lyricist of *The Road to the Isles*. The window bears two inscriptions, one in Gaelic and the other in English. Its central figure is King David, the Psalmist, surrounded by Kenneth

MacLeod's three favourite saints, Columba, Patrick, and Bride.

The chief attraction of Gigha, however, is unquestionably the Achamore Garden, created by Sir James Horlick around his home, Achamore House, built in 1884. In less than a quarter of a century Achamore Garden has become one of the finest of its kind in the British Isles. In 1962 Sir James gave the plant collection of the garden, together with an endowment for its future maintenance, to the National Trust for Scotland. Climatic factors helped the creation of the garden in that Gigha has a medium rainfall of 45 ins in most years and the North Atlantic Drift Current, combined with the influence of the sea, moderates the temperature of the island in winter. Nevertheless shelter belts of trees and windbreaks had to be created before the garden could be developed. Fortunately Sir James brought to the island a lifetime's experience of gardening and garden planning, and today Achamore Garden extends over more than 50 acres and its sign-posted walk covers 2 miles. Within the main garden there are more than 50 separate gardens, to which more are added each year.

For protection the original woodland was extended to the crest of the hills, Sitka spruce and lodgepole pine being planted to supplement the local gorse in providing a windbreak against the Atlantic gales. The guide-book to the garden states that 'it has the air of being made for the plants rather than for human indulgence in them.' Perhaps one could put it this way: Achamore is a gardener's garden, one which provides ideal conditions for inspecting individual plants. It does not pander to those who want to see a mere 'pretty show'.

Rhododendrons are grown in great profusion and in many varieties. However, it is not merely the variety of rhododendrons which has made the garden justly famed, but the quality of cultivation – the soil preparation, and the patient moving around of each plant until the right site has been found. One of the most magnificent of the plants is a *Rh. falconeri*, now 30 years old, 17 ft high and over 20 ft wide, which blossoms into billowy yellow trusses in May. Dwarf rhododendrons are also extensively grown in the more open sections of the garden, some of the best being found in the North Walled Garden.

Azaleas are planted to achieve a formalised perfection in single-colour blocks of white, pale yellow, and orange. There are some magnolias and even camellias in the Walled Garden, as well as stuartia (the deciduous relatives of the camellia), mahonia, viburnum – the Viburnum Garden has an extensive collection of this plant – cotoneasters, hydrangeas, Chilean plants, Australasian plants, and even some plants from Mexico, California, and South Africa. Roses also thrive, despite the fact that they are not normally a success on the west coast of Scotland, and the Walled Garden has a large collection of hybrid musk roses.

Climbing plants such as jasmines, solanums, and honeysuckle flourish on the walls around the Walled Garden. There are many ornamental trees, though most of these have still to reach maturity. And all this is to give only a slight indication of the botanical treasures of Gigha.

The ornithologist will find a great variety of birds on the island, and a number of sacred stones, cairns, and burial places will interest the archaeologist.

See also Cara, Gigalum.

Accommodation: the Gigha Hotel has 3 bedrooms, and also a caravan to hire out. Plans for a new hotel are being discussed.

Places of interest: Achamore Garden is open to the public from April to the end of September.

Post Office.

Books: *The Garden at Achamore House, Isle of Gigha*, by Peter Clough, The National Trust for Scotland, 1970; *The Antiquities of Gigha – a Survey and a Guide*, by R. S. G. Anderson, Galloway Gazette Ltd, 1936.

GIGHAY: *see under* Barra Isles.

GILSAY: *see under* Sound of Harris Islands.

GOMETRA

SITUATION: linked by road bridge to the island of Ulva in the Inner Hebrides, 2 miles S. of the NW. of Mull.
AREA: 2 miles by 1 mile.
POPULATION: 33.
ACCESS: by road from Ulva, reached via the ferry from Mull.

Like nearby Ulva, Gometra was once quite densely populated, and was given over to agriculture and the production of kelp. It is a wild, rugged island, never rising above 500 ft, with a precipitous coastline. There is a private anchorage and quay at the E. end.

Cattle used to be found on Gometra, but the introduction of sheep upset the balance in the grazing land and this, combined with the failure of potato crops in the last century, led to its inhabitants leaving for the mainland.

GRAEMSAY

SITUATION: in Hoy Sound at the NW. approach to Scapa Flow in the Orkneys, 2 miles from Orkney Mainland and 1½ miles from Hoy.
AREA: 250 acres.
POPULATION: 5.
ACCESS: by boat from Orphir or Stromness on Orkney Mainland.

Graemsay's soft, green outlines can easily be identified by the white turrets of two lighthouses, one high and one low. It provides a vantage point for surveying Scapa Flow, with views across to Orkney Mainland and Hoy. The sea can be treacherous in its vicinity, for not only does the tide in Hoy Sound run as fast as in the Firth, but there are many whirlpools.

See also Orkney Islands.

GRALISGEIR: *see under* Sulasgeir.

GREAT CUMBRAE: *see under* Cumbraes, the.

GRIMSAY

SITUATION: 1 mile S. of North Uist and 1 mile NE. of Benbecula in the Outer Hebrides.
AREA: $2\frac{1}{2}$ miles by 1 mile.
POPULATION: 29.
ACCESS: by boat or by causeway from North Uist.

Although a tiny island, Grimsay has a thriving lobster business, dispatching thousands of lobsters each week by air to London and Europe. Visitors are welcome to inspect the ponds and buy if they wish.

There is also a boat-builder on the island, making Viking-shaped lobster boats without the use of power tools. It is even possible to get accommodation here.

Accommodation: limited.
Post Office.

GROAY: *see under* Sound of Harris Islands.

GRUINARD

SITUATION: in Gruinard Bay, 2 miles N. of the Ross and Cromarty mainland.
AREA: 520 acres.
POPULATION: uninhabited.
ACCESS: forbidden to the public.

Situated in one of the most beautiful bays in NW. Scotland, Gruinard is possibly the only island in the British Isles to which public access is rigorously denied at all times. It was requisitioned by the Government in 1942 for secret wartime experimental work to assess the threat from biological warfare. As a result of this work parts of the island were contaminated with anthrax spores, and at the end of the war it became apparent that the island would remain dangerous for many years.

It was therefore purchased from the owner in 1947 by the Ministry of Supply, and is now the property of the Ministry of Defence. It is visited periodically by an inspection team from the Microbiological Establishment, Porton, to assess the virulence of the organism. The island remains contaminated and dangerous, and warning notices are displayed at sixteen carefully sited positions as well as at two points on the mainland. The notices read: 'Gruinard Island. This island is Government property under experiment. The ground is contaminated with anthrax and dangerous. Landing is prohibited. By order 1972.' The date is amended each year.

The terrain of Gruinard ranges from bare rocks and precipitous rock faces to grass and peat bog. The only building remaining on the island is a stone-built shepherd's hut.

GUILLAMON

SITUATION: $1\frac{1}{2}$ miles SE. of Scalpay and $1\frac{1}{2}$ miles N. of the Skye coast.
AREA: 300 yds by 130 yds.
POPULATION: uninhabited.
ACCESS: by boat from Scalpay or Broadford, Skye.

HAAF GRUNEY

SITUATION: 2 miles SE. of Unst in the Shetlands.
AREA: 50 acres.
POPULATION: uninhabited.
ACCESS: by boat from Uyeasound, Unst.

HANDA

SITUATION: 3 miles NW. of Scourie and 1 mile W. of the Sutherland mainland.
AREA: $1\frac{1}{4}$ sq. miles.
POPULATION: 3.
ACCESS: by boat from Scourie or Tarbet village.

A nature reserve run by the Royal Society for the Protection of Birds since 1962, Handa offers special advantages to the ornithologist, not least because of its accessibility. It is a simple boat trip across the Sound of Handa from the tiny village of Tarbet.

Thousands of birds nest on the ledges of the cliffs which, on the NW. side of Handa, rise to nearly 500 ft. On the SE. corner of the island is a stretch of sandy

Handa, the bird sanctuary (J. Allan Cash)

beach. Conditions for studying the bird life at close quarters are not only admirable, but much safer than in many other such reserves where the cliffs present hazards for bird-watchers.

Razorbills, gullemots, kittiwakes, and auks can be found on Handa, and it is said that the peregrine has come here in recent years. At the NW. extremity is the Stack of Handa, a rocky formation which in midsummer is crowded with nesting sea birds. So precipitous is the Stack that it was regarded as unclimbable until in 1877 it was mastered by two men and a boy who had been instructed to reach the summit to destroy the nests of the predatory black-backed gull. It is reliably claimed that the black-backed gull has never returned to the island.

Handa's chief peak is Sithean Mor (406 ft), so named

in Gaelic by the earlier inhabitants of the island because it was reputed to be the home of the fairies. Up to 100 years ago about eight families lived on the island, mainly engaged in crofting. But the Potato Famine of 1845 forced the last of the inhabitants to emigrate to America. Charles St John in his *Tour of Sutherlandshire* described arriving on the deserted island in 1848 and seeing a large, white cat sitting on a small promontory: 'I could not help being struck with the attitude of the poor creature', he wrote, 'as she sat there looking at the sea.' He also mentioned having passed several huts on the island, occupied by starlings. Today some slight traces of old cottages on Handa remain, but even the island's original graveyard is barely discernible.

More interesting are the tiny lochs of Handa, where one can come across red-throated divers. A grant from the Helena Howden Trust has enabled the derelict bothy on Handa to be restored and converted into a headquarters for visiting ornithologists. Water is piped from a nearby loch, and bunks and calor gas facilities for cooking are provided. There is also a small library of reference books.

The Stack still represents a major challenge to rock climbers. When it was first tackled in 1877 a rope was thrown across to the Stack from the nearest cliff point on Handa itself. Today more sophisticated mountaineering equipment is used, but it still requires a rope of about 600 ft to bridge the gap between the Stack and the island.

See also Eddrachillis Islands.

Accommodation: at the bothy. This can be booked only, as a general rule, by members of the RSPB. A small charge is made to cover maintenance expenses, and though blankets and pillows are supplied, visitors must bring their own food, sleeping-bags, etc.

HARMETRAY: *see under* Sound of Harris Islands.

HARRIS: *see* Lewis and Harris.

HASCOSAY

SITUATION: ½ mile E. of Yell and 1 mile W. of Fetlar in the Shetlands.
AREA: 1⅛ sq. miles.
POPULATION: uninhabited.
ACCESS: by boat from Yell or Fetlar.

At one time Hascosay was inhabited, and despite the large area of peat moor, there are still expanses of fertile land here. It forms a natural shelter for Yell's harbour and possesses some fine cliff walks with many caves to explore. There is also an abundance of sea birds.

See also Shetland Isles.

HEISGEIR ISLAND

SITUATION: 12 miles N. of the Monach Isles, 10 miles W. of North Uist in the Outer Hebrides.
AREA: 4 acres.
POPULATION: uninhabited.
ACCESS: by boat from Tighary in North Uist.

The largest of a group known as the Heisgeir Rocks, Heisgeir Island is the only one to possess vegetation. Heisgeir Eagach, a cluster of some eight rocks, is to the SW. of the main island.

All the islands and rocks are the haunts of Atlantic seals which have been almost their only inhabitants, except for the sea birds, from time immemorial. In the past there was much indiscriminate and uncontrolled killing of seals at Heisgeir, but since the Act of 1931 made the killing of grey seals illegal, Heisgeir has become their sanctuary.

Heisgeir Island is highest at the north end, where the

cliffs rise to 120 ft. There is no anchorage in the area, and the only safe landing point is on the eastern hump of the island when the weather is good.

HELLISAY

SITUATION: off Northbay, Barra, in the Outer Hebrides.
AREA: 8 acres.
POPULATION: uninhabited.
ACCESS: from North Bay, Barra.

Sometimes called 'The Isle of the Fairies', Hellisay's sole claim to fame is that according to Gaelic folk-lore it was once inhabited by fairies, just as Fuiary (q.v.) has its fairy washerwoman. Legend has it that the fairies were so numerous that the crofters were very careful not to offend them in any way. If they did they were apt to find their livestock turned loose among the corn, or driven to some remote corner of the island. To placate the fairies the crofters put out bowls of milk for them. In his book *The Enchanted Isles* Alasdair Alpin MacGregor tells of one Angus Macneil who was reported to have actually 'witnessed a faery wedding on his native Hellisay . . . one evening he observed a multitude of The Little People emerge in pairs from an adjacent knoll. So exquisitely attired were they that he hadn't the slightest doubt about their being guests at a faery wedding.'

See also Barra Isles.

HESTAN

SITUATION: at the entrance to Auchencairn Bay, Kirkcudbrightshire, ¾ mile offshore.
AREA: ¼ mile by 120 yds.
POPULATION: 5.
ACCESS: at low tide on foot across a long, narrow spit of gravel; otherwise by boat.

Hestan Island was immortalised by S. R. Crockett in his book *The Raiders*, in which he gave it the name 'Isle Rathin'.

It enjoyed a brief period of fame at a much earlier date (1332) when it became the headquarters of the puppet king, Edward Balliol, after he was crowned at Scone. Edward never gained wide support, but he held sway for some time in Galloway and on Hestan, where today, at the N. end of the island, stands the ruin of the manor from which he issued grants under the Great Seal of Scotland.

Apart from this ruin, Hestan is interesting because of the midden of oyster shells, probably of Mesolithic date, which lies at the top of the steep beach close to a 19th-century farmhouse.

HILDASAY

SITUATION: 3 miles W. of Scalloway on Shetland Mainland.
AREA: 255 acres.
POPULATION: uninhabited.
ACCESS: by boat from Scalloway.

An island capable of modest cultivation, Hildasay had a population of thirty-one in 1851, which had dropped to three by 1871, and increased to thirty by 1891; but at the turn of the century it was uninhabited. There is ample scope here for an environmental study of recent occupation and even of future development.

See also Shetland Isles.

HINBA ISLAND: *see* Eileach an Naoimh.

HIRTA

SITUATION: the principal island of the group of four St Kilda Isles situated about 34 miles WNW. of North Uist in the Outer Hebrides.

AREA: 2½ sq. miles.

POPULATION: none permanently, but Army and National Trust personnel occasionally stay here.

ACCESS: the National Trust for Scotland runs cruises around the Hebrides which take in Hirta.

Rising steeply from the ocean, Hirta's chief landmark is the 1,397 ft peak of Conachair. The coastline bears all the marks of erosion by the sea, and the cliffs are pocked with large caves, some of which extend 300 ft underground. The rocks are of volcanic origin, and include elements of felspar, magnetite, granite, and basalts.

The best landing site is Village Bay on the SE. corner of the island, where there is a small pier. Two hills overlook the bay and behind them is the towering peak of Conachair, down which runs a stream. There is grass right up to the summit of the hills, and on the slopes of Conachair are to be seen sea-pink and camomile. The island is rich in bird life ranging from the gannet and puffin to the razorbill and fulmar. In the past these sea birds were not only a source of food for the inhabitants but, in the case of the fulmar, provided oil for export as well. It is estimated that in the bird-hunting season as

Hirta, chief of the St Kilda Isles, showing the ruined village (Aerofilms)

many as 20,000 birds were killed in a single year on Hirta.

The island also possesses a unique type of wren (*Troglodytes hirtensis*) and this bird was the subject of the Wild Birds Protection Act of 1904, primarily intended to safeguard the St Kilda wren and Leach's fork-tail petrel. The auk was once to be found on the island and its oil was used for lamps, but today it is totally extinct in this area. Hirta has the world's largest gannetry, numbering about 17,000 breeding pairs. Kittiwakes, shearwaters, and storm petrels are also to be found.

The island's name, Hirt or Hirta, comes from the Gaelic *H-Iar-Tir*, meaning 'westland'. 'Kilda' may be a corruption of this. There is no known saint of the name Kilda; but it may be derived from the Norse *kelda*, 'a well'. The island was named 'Kilda' on a map of 1558, and it is said that a Dutch map-maker added the prefix 'St' much later. The population of the island was at its peak in 1697 when one Martin Martin visited it and recorded that there were 180 people living on the shores of Village Bay.

The MacLeod clan owned the island for hundreds of years, and Murray of Broughton in his report on conditions in the Highlands in 1746 said of the St Kildans: 'Their Chief [the MacLeod chief] is their God, their everything.' Usually the MacLeod of MacLeod deputed one of his family to act as steward of Hirta. The latter would visit the island annually and collect the rents, which were paid in cloth, feathers, wool, butter, cheese, cows, fowls, oil, and barley. Any surplus funds were used to buy goods for the islanders from the mainland.

There was a rough and ready form of administration on the island – the St Kilda 'Parliament', which was composed of all the males on the island and which had power over all members of the community. This body had no chairman and there appear to have been no rules governing it.

From the beginning of the 18th century the population of Hirta gradually declined, so that even as long ago as 1875 there was a plan to evacuate all the inhabitants to Canada. Nothing came of this, but by 1930 there were only thirty-seven people left on the island, and Mr T. B. W. Ramsay, MP for the Western Isles, wrote to the Secretary of State for Scotland asking for help for 'the people of the island who are in such dire straits as regards food supplies.'

This letter was followed on 10 May 1930 by a petition from the St Kildans to the Secretary of State requesting him 'to assist us all to leave the island this year and to find homes and occupations for us on the mainland.' The evacuation took place on 29 August 1930, when HMS *Harebell* removed the inhabitants to Oban.

The National Trust for Scotland, which now owns the island, has done much to preserve what is left of old Hirta, keeping some of the cottages in good repair, while the Nature Conservancy, to which it is in part leased, has conducted an ecological survey of the island. But the face of Hirta changed rapidly when in 1957 the British Government set up a rocket-firing base at South Uist, and the Ministry of Defence sent military personnel, together with a few Hebridean civilians, to the island to track guided missiles fired from South Uist. In recent years the pier has been extended, power and radar stations established, and new living quarters built at the military camp.

Attempts to bring organised religion to Hirta in the 18th century was somewhat primitive, and services were held in a barn. A combined church and manse was erected in the early part of the 19th century. The island then came under the domination of a somewhat harsh Calvinist, the Revd John MacKay, who even censored the children's books. The Free Church pastor of the Hirtans in this era was even called upon to adjudicate in breach of promise cases, and it is recorded that one pastor, the Revd Neil MacKenzie, fined the defaulting male '100 full-grown fulmars, 50 young solan geese and a hair rope.' The last-named was regarded as the most valuable item of all as it was used for bird-catching.

In education the island lagged sadly behind the mainland for many years. It was not until the summer of 1884 that the first schoolmaster arrived on Hirta. In the *St Kilda School Log Book* there is an interesting entry for 10 December 1918 which says: 'School re-opened today. No school has been held since the island was bombarded [by a German submarine] on 15 May.

During most of that time the school has been occupied by a party from the naval station while their own premises have been repaired.'

See also St Kilda Isles.

Books: *St Kilda and other Hebridean Outliers*, by Francis Thompson, David & Charles, 1970; 'Tour in the Western Isles, including St Kilda, in 1799', by Lord Brougham, in *The Life and Times of Henry, Lord Brougham*, vol. ii, National Trust for Scotland, 1871; *The Life and Death of St Kilda*, by T. Steel, National Trust for Scotland, 1965; *Edge of the World*, by Charles Maclean, Tom Stacey, 1972.

HOLM OF PAPA: *see under* Papa Westray.

HOLY ISLAND

SITUATION: ¾ mile w. of Arran, off Lamlash Bay.
AREA: 2 miles by ¾ mile.
POPULATION: 25.
ACCESS: by rowing boat in good weather, or by motor-boat from Lamlash.

Forbidding, beetling cliffs of igneous rock dominate the eastern and western sides of Holy Island, which at one time was joined to the larger island of Arran (q.v.). It is a great hulk of an island, rising to 1,030 ft, inhabited mainly by goats, and there are far fewer of these in recent years. It would seem that the original Arran goats disappeared from the main island at the beginning of the last century, and are now only to be found on Holy Island.

There is some interesting bird life here, notably the fierce peregrine, which inhabits the higher cliffs, and sometimes the kestrel, merlin, and meadow pipit. The shag has also been known to nest on Holy Island.

Dean Monro, after visiting Arran in 1549, described Holy Island as the 'yle of Molass'. The meaning of the phrase has never been absolutely solved, but it is said to refer to St Molas, who was born in AD 566 and is believed to have lived in a cell on the island. Whatever the truth of the legend, the cave of St Molas is still pointed out to visitors to the island, and some of the carvings and writings on the walls of the cave do indeed suggest that it was a place of pilgrimage centuries ago. There was a chapel there at one time, and writers have mentioned seeing its ruins about the middle of the 18th century. But more interesting to archaeologists are the Viking names carved on the walls of the cave of St Molas. These probably date from round about the time of the Battle of Largs (1263), when the Scots finally routed the invaders from Norway. As a result of this Arran became part of the Kingdom of Scotland. It is known that King Hakon of Norway and his fleet anchored in Lamlash Bay that same year. No doubt at that time some of the Vikings went ashore on Holy Island and carved their names on the walls of the cave.

Holy Island was sold to Mr Stewart Huston, an American with Scottish ancestral associations, some few years ago.

Books: *The Isle of Arran*, by Robert McLellan, David & Charles, 1970; 'Viking Burials', 'Fortified and Domestic Sites', 'An Irish-Celtic Monastery', 'The King's Cave', and 'The Holy Isle', by J. Balfour in *The Book of Arran*, vol. 1, Arran Society of Glasgow.

HORSE ISLAND (Ardrossan, Ayrshire)

SITUATION: 2 miles W. of Ardrossan, Ayrshire.
AREA: 5 acres.
POPULATION: uninhabited.
ACCESS: by boat from Ardrossan.

The island has been used for grazing, and at one time ponies were kept here.

HORSE ISLAND (Muck): *see* Eilean Nan Each.

HORSE ISLAND (Summer Isles)

SITUATION: one of the Summer Isles, 2 miles W. of the Ross and Cromarty mainland.
AREA: 350 acres.
POPULATION: uninhabited.
ACCESS: by boat from Ullapool.

Horse Island is noted for its herd of wild goats, and for having the best collection of bird life in all the Summer Isles (q.v.). It was at one time inhabited.

HOUSAY: *see under* Skerries (Shetlands), the.

HOY

SITUATION: in the SW. corner of the Orkney Islands group, 10 miles N. of Mell Head in Caithness.
AREA: 14 miles by 6 miles.
POPULATION: 438 (1970).
ACCESS: by coal tender from Scrabster, Orkney Mainland.

Second largest island in the Orkneys (q.v.), Hoy is also scenically the most spectacular, especially for those who enjoy massive piles of rock. Its chief landmark is the famous Old Man of Hoy, a solitary peak of rock columns rising to a height of 450 ft, jutting out from the sea's edge at the end of a peninsula.

The Old Man of Hoy is a delight equally to the hardiest and most ambitious of rock climbers and to geologists. For the former it is one of the toughest climbing challenges presented in the British Isles, while for the latter the basalt lavas deposited on an ancient land surface which form a pedestal beneath the Old Man are of special interest. Hoy is built almost entirely on upper old red sandstones not to be found elsewhere in Orkney: they are a distinctive orange-red colour.

There is a good road on the island, but the best views are to be obtained by walking along the western cliffs towards the Old Man of Hoy. Another place worth visiting is the Dwarfie Stone, which is above the col on the road from Linksness Pier to Rackwick. It is the subject of many local legends, and probably is the oldest of all the Neolithic tombs in the Orkneys. A vast sandstone rock, 28 ft by 14 ft, has been hollowed out to create a corridor with two chambers. Close to the stone is another large stone which once sealed the entrance. The Dwarfie Stone was identified in 1935 as a cut-out variant of Maeshowe by Mr C. S. T. Calder, of the Royal Commission on Ancient Monuments of Scotland. A century previously it had inspired the romantic poets and novelists, and the Dwarfie Stone is featured in Sir Walter

The Old Man of Hoy (Aerofilms)

Scott's *Pirate*.

In 1851 Hoy had a population of 1,551. Today most of the inhabitants live in Melsetter, Brims, and South Walls. There is a school, and some good arable land.

The Forestry Commission in an experiment in growing trees in gale-swept locations has established a coniferous plantation at Hoy Lodge at the NE. end of Hoy, but the trees, which were planted in 1953, are still stunted in growth.

Hoy is, with North Ronaldsay (q.v.), the only estate island left in the Orkneys, and farming is mainly confined to the low-lying S. end of the island. Worth visiting in the N. is Ward Hill (1,570 ft) and St John's Head in the W., 2 miles N. of the Old Man of Hoy, should not be missed. This headland towers some 1,140 ft above the sea.

Accommodation: some cottagers take in visitors.

Books: *Orkney*, by Patrick Bailey, David & Charles, 1971; *Poems New and Selected*, by George Mackay Brown, Hogarth Press, 1971; *Eday and Hoy: a Development Survey*, by Ronald Miller and Susan Luther-Davis, Dept. of Geography, University of Glasgow, 1970.

HUNDA

SITUATION: 1 mile W. of Burray in the Orkney Islands.
AREA: 80 acres.
POPULATION: 3.
ACCESS: by boat from South Ronaldsay.

The island has some interesting bird life.

See also Orkney Islands.

HUNEY

SITUATION: ½ mile E. of Unst in the Shetlands.
AREA: 56 acres.
POPULATION: uninhabited.
ACCESS: by boat from Baltasound, Unst.

Worth a visit for the bird life and some of the rarer plants of the Shetlands (q.v.).

INCHCOLM

SITUATION: in the Firth of Forth, 1½ miles S. of Aberdour, Fife.

AREA: 1 mile by ½ mile.

POPULATION: varies with the seasons, but about 4.

ACCESS: by boat from Aberdour, or occasionally in summer by steamer from Granton on the S. shore of the Firth of Forth, close to Edinburgh.

Inchcolm is the Gaelic for 'the isle of Colum', and it received its name from the fact that a Columban missionary or monk set up a cell there.

Legend has it that this hermit settled on the island in the 12th century, and because of the hospitality he showed to Alexander I of Scotland when the latter was forced to seek shelter there after a storm at sea, the King ordered that an Augustinian abbey be built on Inchcolm as a thank-offering. One can see the well-preserved ruins of the Abbey of St Columba, originally built in 1123, though the present ruins are mainly those of the 13th-century reconstruction. Visitors are shown a small cell that is said to be that of the original hermit.

The reputation of this abbey for possessing an aura of holiness must have extended far beyond the borders of Scotland, for Shakespeare refers to it in *Macbeth*. In Act I, scene ii, Ross reports:

> That now
> Sweno, the Norways' King, craves composition,
> Nor would we deign him burial of his men
> Till he disbursed, at St Colme's-inch,
> Ten thousand dollars to our general use.

As a burial site the island was regarded as being 'an island close to heaven', and much coveted on this account. But after the Reformation the buildings were used as a residence of the Earls of Moray. Late in the last century Inchcolm became the property of the nation. It is a

Inchcolm Island; the Abbey (Scottish Tourist Board)

verdant island with excellent grazing land and splendid views of both banks of the Firth of Forth.

INCHKEITH

SITUATION: in the Firth of Forth, 4 miles SSE. of Burntisland and 4 miles N. of Leith.
AREA: 1 mile long and $\frac{1}{2}$ mile wide at its N. extremity, tapering to a point at its S. tip.
POPULATION: 33.
ACCESS: by boat from Leith.

This island takes its name from the family of Keith, who held the office of Hereditary Grand Marischal of Scotland, but its chief claim to fame is as a military outpost guarding the Firth of Forth.

Its importance in this respect was first recognised by Mary of Lorraine who, as Queen Regent, invited her fellow-countrymen from France to defend it on behalf of Scotland in the 16th century. It was garrisoned by the French for many years, until the time of Mary, Queen of Scots. The ancient fortress of Inchkeith has long since disappeared, and modern gun emplacements and radio equipment have taken its place. Inchkeith has been a valuable minor military establishment in two world wars, and its lighthouse is a landmark for miles around.

Permission to visit the island can be obtained in Leith. There are magnificent views across to Edinburgh, but otherwise there is little to see there.

INCH KENNETH

SITUATION: 1 mile S. of the entrance to Loch Na Keal, off the W. coast of the isle of Mull.

AREA: 200 acres.

POPULATION: 12.

ACCESS: no regular service, but the owner of the island has a private ferry boat which runs to Mull.

This island takes its name from Kenneth, a contemporary of St Columba, whom he is reputed to have saved from drowning. Kenneth was Abbot of Achabo in Ireland. In the 16th century Inch Kenneth was the property of the Prioress of Iona. There are the remains of a 12th-century chapel on the island today, measuring 40 ft by 30 ft, and various tombstones of some antiquity.

When Dr Johnson visited Inch Kenneth in 1773 with

his friend Boswell, it was owned by Sir Alan Maclean, chief of Clan Maclean, and the often hard-to-please doctor was moved to describe Inch Kenneth as 'a pretty little island'. It is still a fertile island, with a flat surface which never rises higher than 160 ft. Despite the low-lying cliffs there is a good landing place.

Until a few years ago the island was the property of Jessica Mitford, the authoress. In 1965 she sold it to Dr Andrew Barlow, a London consultant, for £25,000. He lives in the large Victorian mansion, and apart from supervising farming on the island and keeping cattle, sheep, and hens, he breeds Shetland ponies.

It is a delightful retreat which in a small area offers scope for good fishing and contains, apart from the ruined chapel, a farmhouse, a secret cave (said to have been a hermit's cell), and the cottage where Dr Johnson and Boswell stayed.

INCHMARNOCK

SITUATION: 1½ miles W. of the Isle of Bute, and 5 miles E. of the Mull of Kintyre.
AREA: 2½ miles by ¾ mile.
POPULATION: 13.
ACCESS: by boat from Bute, but often more easily from Kintyre.

Inchmarnock is part of the county of Bute, and is much more like an island than the Isle of Bute, to which it belongs. It is a long, low-lying island suitable for grazing, and provides some splendid views across to Kintyre and Bute. It has the remains of a chapel dedicated to St Marnock.

INCHMICKERY

SITUATION: midway between the N. and S. banks of the Firth of Forth, $2\frac{3}{4}$ miles N. of Cramond and 4 miles SE. of Burntisland.
AREA: 30 acres.
POPULATION: uninhabited.
ACCESS: by boat from Cramond.

Inchmickery lies slightly to the S. of a rocky islet known as Oxcars, and has a certain amount of pasture land of a poor kind. It rises to a height of only 48 ft, and is conspicuous only because of the narrow ledge off its S. extremity on which stands the 14-ft-high Mickery Stone.

IONA

SITUATION: $\frac{3}{4}$ mile off the SW. of the Isle of Mull in the Inner Hebrides.
AREA: 3 miles long and $1\frac{1}{2}$ miles wide.
POPULATION: 234.
ACCESS: by ferry from Fionphort; some steamers from Oban also call at Iona.

This is a small island with a rich history out of all proportion to its size. It was here in AD 563 that St Columba arrived from Ireland, and from here that he launched his crusade to bring Christianity to Scotland. For this reason it has been a centre of pilgrimage over the centuries: 'that man is little to be envied whose . . . piety would not grow warmer among the ruins of Iona', declared Dr Johnson.

Columba and his twelve disciples landed on the S. tip of the island by coracle: it is recorded in the Woodrow

Iona and its Cathedral (Aerofilms)

MS in the Advocates' Library, Edinburgh, that St Columba's coracle measured 60 ft in length. A follower of St Patrick, Columba established a monastery on Iona, and this became a centre not only for missionaries, but scholars as well. The original buildings here, of daub and wattle, have long since disappeared. After Columba's death Iona was ravaged by the Norsemen, and it was not until the 7th century that the remaining monks built an abbey which again suffered from the invading pagans.

In the 12th century Somerled, a chieftain from Argyll, married a member of the Norwegian royal family and became the owner of the island. His son founded a Benedictine community on Iona in 1203, and a few years later a convent was established there. The ruins of these buildings still remain, covered with flowers in summertime.

For many years Iona was the burial ground of Celtic kings – Irish and Scots and, it is said, French and Norwegian as well. Forty-eight Scottish kings are buried here, including Macbeth and Kenneth MacAlpine, the first Celtic king of Scotland, together with many clan chieftains of the neighbouring islands. Under the last Abbot of Iona, Mackinnon, Iona was the centre for the bishopric of the Isles. With the Reformation the island suffered again, and of the 360 crosses said to have been erected there only three now remain. Some of the monks fled with books and records that eventually found their way to the Vatican.

In St Columba's time it was a tradition that neither women nor cattle were allowed to live on the island – an old Ionan saying, attributed to Columba, is:

> Where there's a cow there's a woman;
> And where there's a woman there's mischief.

The arrival of the nuns put an end to that unwritten rule.

After the monastery was suppressed in the 16th century the island was taken over by the Macleans of Duart until in 1693 Iona was handed over to the Duke of Argyll. The eighth Duke gave the ruined abbey buildings to the Church of Scotland in 1899 'for the use of all Christian denominations'. The first service in the abbey was held in 1910, but the buildings were not finally restored and reconstructed until 1938, when the Iona Community, a body of voluntary workers under the Revd George MacLeod, undertook the work. Services are now held there regularly. The restoration of the buildings was supervised by the Scottish architect Ian Lindsay.

The island, however, has many interesting features besides its abbey. Dun Hill, its highest point (332 ft), is well worth a visit, and there is a cairn erected in memory of St Columba near its summit. The island's sands are a dazzling white, and on its S. and W. shores are to be found the exquisitely-coloured Ionan pebbles. These are only an inch or two in length and are dark green and greenish-yellow. Marble is also to be found on the island.

In the middle of the last century Iona had a population of 500, but it has fallen steadily since that time. Today the inhabitants are mostly concentrated in the little village which is connected by ferry with Fionphort in Mull. No cars are allowed on the island.

Accommodation: there are 2 hotels on Iona, and some of the islanders take visitors.

Books: *St Columba of Iona*, by Lucy Menzies, 1935; *Iona Past and Present*, by T. Ritchie, Highland Home Industries Ltd, 1945.

IOSA: *see under* Ascrib Islands.

ISLAY

SITUATION: in the Inner Hebrides, separated from the Isle of Jura by the 1-mile-wide Sound of Islay, 18 miles W. of Kintyre.
AREA: 401 sq. miles.
POPULATION: 4,200 (1970).
ACCESS: by air from Glasgow or Campbeltown to Port Ellen, or by car ferryboat service from West Loch Tarbert to Port Ellen or Port Askaig.

Islay is the most southerly of the Hebrides, famed for its whisky production: it has no fewer than eight distilleries. As you explore the island you can always tell when you are in the proximity of a distillery by the aroma of smouldering peat from the piles kept by the distilleries.

Distilling is one of the prime occupations on the island, the others being principally farming and fishing. The visitor is well catered for in accommodation and in facilities and recreations. There are shooting, fishing and sea-angling, and an eighteen-hole golf course at Machrie.

Islay: Port Ellen (British Tourist Authority)

Bird-watchers will be interested to know that no fewer than ninety-six species have been recorded on the island in two days' winter watching. Islay is one of the few places in the British Isles where the chough breeds. Roads are good for motoring, and bus services operate.

The western side of the island is very rugged, and

large herds of red deer can be seen in this area. But on the whole Islay is a fertile, low-lying island, with Bowmore (its chief settlement) almost at its centre. Bicycles can be hired here to make tours of the island, and indeed this is the most rewarding method of exploring the place.

While Bowmore is the capital of Islay, Port Ellen is the largest village and port of call for ships, though Port Askaig, on the E. side facing Jura, is also used by ships. Bridgend is the chief road junction.

Islay's history dates back to Neolithic times, and at one time the island had close links with Ireland. After the end of Norwegian rule in the 13th century the island became the headquarters of the Lordship of the Isles that made the Macdonalds supreme. They established themselves in Dunyveg Castle, which is today a ruin. The stonework of the keep is, however, still more or less intact. The Macdonalds also had another castle, situated on an islet in Finlaggan Loch: in dry weather one can walk out to inspect its ruins and see the rather fine carved gravestones.

In 1598 the Macleans attacked the Macdonalds of Islay on the mudflats of the N. coast near Loch Gruinart. They came in vastly superior numbers and their strategy was to capture the peninsula on the W. side of Islay cut off by the narrow strip of land between Loch Gruinart and Loch Indaal. But the Macdonalds triumphed and drove the invaders back to Mull.

Not to be missed is Kildalton Chapel on the SE. tip of the island. This is being restored, and in the graveyard are some exquisitely carved stone crosses, one of which, dating back to the 8th century, is one of the finest examples of early Celtic work. This is the celebrated Kildalton Cross, 8 ft high and mounted on stone steps, its carving representing Abraham about to sacrifice his son Isaac as instructed by God.

Kildalton Castle nearby has fine gardens laid out with many unusual plants and shrubs that give them a sub-tropical flavour. There are more woods in this part of Islay than one normally finds in the Hebrides.

Though mainly low-lying, Islay has an area of wild upland on the E. side. Its chief peak is Beinn Bheigeir (1,609 ft), from the summit of which views of the nearby islands, especially Jura (q.v.), are obtained. There is a lighthouse on a steep cliff at the S. end of the Sound of Islay.

The Sound can be beautiful – water which changes in colour from deep blue to green – but is decidedly treacherous and has navigational hazards. Only skilled yachtsmen should sail in and around these water. Bird-watchers will find that Bridgend is the best centre for them: there is a hotel here where advice is given, and the proprietor keeps a book to enter sightings made by visitors. Apart from the chough, already mentioned, snow geese, northern divers, and sooty sheerwaters have been found on the island.

Accommodation: there are hotels at all the largest villages, including Port Ellen, Bowmore, Bridgend, and Port Askaig. The Bridgend Hotel is recommended for ornithologists. It is as well to book accommodation in advance for high summer, and lists are available from the Scottish Tourist Board, 2 Rutland Place, West End, Edinburgh 1.

Books: *Rambles in the Hebrides*, by Roger A. Redfern, Robert Hale, 1966; *The Hebrides and Their Legends*, by O. F. Swire, Collins, 1964.

ISLE MARTIN

SITUATION: one of the Summer Isles, 1 mile NW. of the Ross and Cromarty mainland at the entrance to Loch Broom.
AREA: 600 acres.
POPULATION: uninhabited.
ACCESS: by boat from Ullapool.

Like Tanera More (q.v.) and Ristol (q.v.), Isle Martin had a herring-curing station in the late 18th century, linked with the Ullapool herring industry. It is an attractive island with varied bird life.

See also Summer Isles.

ISLE OF EWE: *see* Ewe.

ORONSAY (North Uist)

SITUATION: 1 mile N. of North Uist in the Outer Hebrides.
AREA: 30 acres.
POPULATION: 7.
ACCESS: by boat from North Uist.

ISLE ORNSAY (Skye)

SITUATION: just offshore from the village of Isleornsay on Skye, in the Sound of Sleat, facing Loch Hourn on the Scottish mainland and 9 miles SE. of Broadford.
AREA: 25 acres.
POPULATION: 4.
ACCESS: by boat from Isleornsay.

This isle takes its name from the Columban saint, Oran, to whom a chapel was dedicated here – the remains can still be seen. There is a lighthouse 63 ft high with a fixed light 58 ft above sea level which can be seen for 12 miles.

ISLES OF FLEET

SITUATION: off the approach to Fleet Bay, Kirkcudbrightshire, ½ mile SW. of Knockbrex.
TOTAL AREA: 30 acres.
POPULATION: uninhabited.
ACCESS: by boat from Fleet Bay.

The Isles of Fleet comprise the three islands of Ardwall, Barlocco, and Murray's Isle, all lying off some of the finest scenery on the Galloway coast.

The largest of the three is Ardwall, ¼ mile from the mainland, to which it is joined at low tide by a narrow, rocky strip. The island measures 4 furlongs by 2½ furlongs, and is 109 ft high at one point. Ardwall is leased by Cordie Co. Ltd from the owner, Mr W. J. McCulloch, of Gatehouse of Fleet, a member of an old Gallovidian family in whose possession the island has been for many years. Its chief feature is an unusual and fairly well preserved *broch*, 30 ft in diameter, with walls 13 ft thick. It has an entrance on the seaward side and another facing inland.

Barlocco is owned by Mr Andrew Brown, of Roberton, Borgue, Kirkcudbrightshire. Like Ardwall it is a small peninsula, isolated from the mainland only at high water. It measures 2½ furlongs by 1½ furlongs, and rises 35 ft above sea level at its highest point.

Murray's Isle belongs to the National Trust for Scotland.

ISLES OF THE SEA, THE: *see* Garvellochs, the.

JURA

SITUATION: 6 miles W. of the Argyll coast, in the Inner Hebrides.
AREA: 141 sq. miles.
POPULATION: 250 (1970).
ACCESS: by air via Islay (BEA from Glasgow) and hired car and ferry; or by boat from West Loch Tarbert to Craighouse. Also car ferry service in summer, via Islay.

One of the most inaccessible islands in the Inner Hebrides despite its relatively short distance from civilisation, Jura is in complete contrast to nearby Islay, having only a twentieth of the latter's population and being wild, rugged and mountainous. Almost split in two by Loch Tarbert, which joins the sea on its W. side, Jura is separated from Scarba (q.v.) in the N. by the perilous Corrievreckan whirlpool, and from Islay in the S. by the narrow Sound of Islay.

It is an island for mountain-lovers, for the three peaks, known as the Paps of Jura, offer ample scope for climbing. The highest of these is 2,569 ft (Beinn an Oir, or the 'Hill of Gold'). From them magnificent views stretch from the Isle of Man to the Outer Hebrides on a clear day. Beinn Siantaidh (the 'Hallowed Mountain') is 2,477 ft, and Beinn-a-Chaolais ('Mountain of the Firth') is 2,407 ft.

Jura (its name comes from the Norse *Deoroe*, meaning 'Deer Isle') is essentially an island for those who like solitude and a kind of lonely grandeur: for this reason it never attracts so many visitors as nearby Islay. There is no public transport on the island. Craighouse is the main port, and Feolin is the pier for Islay. Easily the best way of seeing Jura is the hard way – on foot, and well shod at that, in order to negotiate the largely mountainous terrain.

There are thousands of red deer on the island, mainly in Ardlussa Forest to the N. and Jura Forest to the S. Jura itself is today largely divided up into sporting estates. The community is mainly engaged in small-scale

Paps of Jura and the Sound of Islay (Scottish Tourist Board)

farming, most of the population being centred on the E. coast and in the vicinity of the village of Craighouse.

At the N. tip of the island, close to Ardlussa, is the grave of Mary MacCrain, bearing this inscription:

> died in 1856, aged 128,
> descendant of Gillouir MacCrain,
> who kept a Hundred & Eighty Christmasses
> in his own house and who died in the
> reign of Charles I.

Jura has a river, the Corran, in which there is said to be good fishing. There are some raised beaches and caves on the W. side of the island which are worth exploring.

Accommodation: there is a hotel at Craighouse, and several houses take in visitors.

Book: *Rambles in the Hebrides*, by Roger A. Redfern, Robert Hale, 1966.

KELLIGRAY: *see under* Sound of Harris Islands.

KERRERA

SITUATION: separated from the mainland of Argyll by Kerrera Sound, and at its most northerly point ¾ mile from Oban.
AREA: 5 miles by 1–2 miles.
POPULATION: 103.
ACCESS: by boat from Oban.

Kerrera makes a natural breakwater to protect the harbour of Oban. On the N. end of the island is an obelisk commemorating David Hutcheson, founder of the West Coast steamship services.

The most interesting feature of the island is at the S. end – Gylen Castle, an ancient fort of Norwegian origin which was for centuries a redoubt for the MacDougalls. King Alexander II of Scotland died on Kerrera in 1249 when he was about to mount an offensive against the Norsemen in the Hebrides.

Kerrera has high ground in the centre and there is a road from Balliemore, the chief settlement, to Gylen Castle.

KISIMUL

SITUATION: in Castlebay Harbour, Barra, in the Outer Hebrides.

Kisimul is so tiny that it would not warrant any mention but for the fact that it is the site of Kisimul Castle, the ancient stronghold of the Clan Macneil of Barra. Indeed, looking at the islet from the shores of Barra, one cannot see anything resembling an island, but merely a castle that seems to rise out of the sea. For years it was a ruin, the home of the sea-otter and of birds, but in recent times it has been restored and made habitable by Mr Robert Lister Macneil, the present chief of the clan.

See also Barra Isles.

LADY ISLE

SITUATION: $2\frac{1}{2}$ miles W. of Troon, Ayrshire.
AREA: 10 acres.
POPULATION: uninhabited.
ACCESS: by boat from Troon by special arrangement.

This island, with its lighthouse in Troon Bay, is rocky and uninviting to all except bird-watchers. Roseate and other terns breed here, and the island comes under the Scottish Society for the Protection of Wild Birds. It is a prominent landmark for several miles around. Permission to visit the island has to be obtained in Troon.

LAMBA

SITUATION: in Yell Sound between Yell and Shetland Mainland.
AREA: 119 acres.
POPULATION: uninhabited.
ACCESS: by boat from Shetland Mainland.

This island is said once to have been the home of a reputed witch who made a living by selling 'favourable winds' to seamen. Sir Walter Scott is believed to have consulted her in 1814, though by this time she was living in Stromness.

See also Shetland Isles.

LAMB ISLAND: *see* Craigleith.

LAMB ISLET: *see under* Craigleith.

LEWIS AND HARRIS

SITUATION: in the Outer Hebrides, 24 miles NW. of Skye and 34 miles W. of the Scottish mainland.
AREA: 770 sq. miles.
POPULATION: 30,000.
ACCESS: by air from Glasgow, or by boat from Mallaig or Kyle of Lochalsh, on the Scottish mainland.

This northernmost isle of the Outer Hebrides is also the largest in the group. The island contains two districts: the northern and larger part, Lewis, belongs to Ross and Cromarty, while Harris, the southern section, is part of Inverness-shire. From the cliffs which form the Butt of Lewis in the N. it is possible to drive S. through Stornoway, the only town on Lewis, and Tarbert, the chief port of Harris, to Rodel at the S. end of the island 60 miles away.

Lewis consists largely of flat moorland covered by peat and heather. The majority of the population live in Stornoway or on or near the coast, but considerable efforts have been made to reclaim the moorland for cultivation. This process, begun during the Napoleonic

Ploughing in the Isle of Harris (J. Allan Cash)

Wars, has received a new impetus since World War II. Between 1959 and 1964 some 10,000 acres were reclaimed with the aid of modern fertilisers, though cultivation is still based on the original fertilisation device of mixing the soil with powdered seashells. The chief crops are barley and potatoes. Otherwise the inhabitants are engaged in cloth weaving on a cottage industry basis, or fishing.

Some fifty years ago Lord Leverhulme found himself involved in heavy financial losses in trying to make Lewis and Harris pay, and as a result put the island up for auction in 1925. He is said to have spent more than a million pounds upon a scheme for developing Lewis and

overleaf: *Isle of Lewis; the Standing Stones at Callanish (British Tourist Authority)*

its adjacent islands, yet after his death one estate of 56,000 acres was sold for only £500. He had wanted to make Lewis a great centre for the fishing industry, and he told the people that 'acre for acre, the sea around you contains twenty times more food than your land can produce.' Finally, in September 1923, he summoned a meeting in Stornoway and told the people: 'I give you the island of Lewis.' Yet the people actually refused the gift, though Stornoway accepted the stately castle and grounds.

The most picturesque scenery is to be found on Harris, with its mountainous northern section rising to 2,000 ft in many places and reaching a peak at Clisham (2,622 ft). From here on a clear day one can see across 38 miles of water to the Flannan Isles in the NW. and St Kilda lighthouse to the SW. The Forest of Harris is the only woodland area on the island. To the W. sandy beaches stretch down to Rodel, where the growth of grass and flowers presents a striking contrast to the barren N. This picturesque village, sheltered by the hills to the N., is overlooked by an ancient cruciform church decorated with carvings of a kind one would not expect to find on a building dedicated to religious purposes. Its churchyard contains the grave of Donald MacLeod of Berneray, who fought for 'Bonnie Prince Charlie' at Falkirk and is reputed to have had nine children after his third marriage at the age of seventy-five.

The E. coast provides a sharp and rocky contrast to this. Above and below the road grey, black, and silver-white rocks, smoothly rounded, reach up to the sky and down to the shore. Here are to be found the 'lazy beds', small catches of soil between the rocks, which are cultivated by the crofters.

Lewis and Harris provide a striking example of Gaelic culture preserving its heritage while adapting itself to meet the demands of the modern world. Having survived the attempted colonisation of the islands by the Fife lairds of the 16th century, Cromwell's army and, more recently, Lord Leverhulme's well-intentioned but possibly misguided attempt to modernise the island, the people have shown themselves capable of adapting to future trends. Many young islanders, having made their way in the outer world, have returned armed with new techniques with which to stimulate the economy of the island.

The new tweedmill at Stornoway is an instance of modern technical facilities co-operating with, rather than superseding, the peasant industry in the crofts to produce the famous Harris tweeds. The castle at Stornoway is now a technical college teaching navigation, engineering, and weaving. The Nicholson Institute, founded in the last century, is a further example of the people's desire to educate and help themselves. It exists to help young scholars bridge the gap between the primary school and university.

The people of the island pride themselves on their self-education and their hospitality. They are courteous, but strong in their faith, whether it be in their Presbyterian teachings or their island heritage. An instance of the effectiveness of this combination occurred in 1960 when, led by the Revd Kenneth Macrae, a Free Kirk minister, the islanders resisted a Government scheme to build an airport on the Eye Isthmus, which was also to act as a nuclear base. Perhaps the most strikingly apparent manifestation of the individualism of the people is the bright painting of the exteriors of their houses. The gay colours reflect the Norse influence, particularly in the N.

Apart from the magnificent walks and views, particularly on Harris, the island is famous for its standing stones. Those at Callanish especially are worth a visit. There is a cairn 40 ft in diameter, within which is a miniature cairn in three small divisions, one with a monolith 18 ft high at one end. From this circle single rows of stones radiate E. and W., and double rows N. and S. These stones, thirty-nine in all, mark the graves of ancient warriors. Glen Laxdale in SW. Harris also contains funeral cairns.

It is now possible to leave London at 7.00 am and to

sit down for lunch in Stornoway the same day. On the other hand it is rarely possible to get an evening meal after about 8.00 pm.

The predominant sport is fishing, both sea and fresh-water, though on Sundays this is generally taboo. Within a radius of 5 miles from Stornoway there are five lochs which may be fished free of charge. Many others are free, but some belong to various trusts and a charge is made. On Loch Clachan, for example, it is £5 a day, including a ghillie. Good fishing can also be enjoyed on Harris, notably for salmon and sea-trout, but much of it is strictly preserved and normally let out for a minimum period of one week.

There is a good golf course only a few minutes from Stornoway, and water-skiing facilities are also provided.

Some 16 miles from Stornoway is Dun Carloway Broch, an Iron Age fort – one of the best-preserved in the Hebrides. Equally interesting is the Black House, situated at Arnol, now a museum but previously a dwelling place.

Accommodation: there are excellent hotels at Stornoway, Tarbert, and Rodel.

Post Office and banks: facilities are available.

Books: *Rambles in the Hebrides*, by Roger A. Redfern, Robert Hale, 1966; *Lonely Isles, being an account of Several Voyages to the Hebrides and Shetlands*, by R. Svensson, Batsford, 1955.

LINGA (Muckle Roe, Shetlands)

SITUATION: between Muckle Roe and Shetland Mainland.
AREA: 170 acres.
POPULATION: uninhabited.
ACCESS: by boat from Shetland Mainland.

LINGA (Unst, Shetlands)

SITUATION: midway between Unst and Yell in the Shetlands, ½ mile E. of Yell.
AREA: 122 acres.
POPULATION: uninhabited.
ACCESS: by boat from Yell.

This sheltered island in Yell Sound was inhabited as late as the middle of the 19th century. It has picturesque views of the other Shetland Isles (q.v.), and is 1½ miles long from N. to S.

LINGA (Vaila, Shetlands)

SITUATION: between Vaila island and Walls, Shetland Mainland.
AREA: 60 acres.
POPULATION: uninhabited.
ACCESS: by boat from Walls.

LINGAY (Eriskay)

SITUATION: 2 miles W. of Eriskay island and 2¼ miles S. of South Uist in the Outer Hebrides.
AREA: 40 acres.
POPULATION: uninhabited.
ACCESS: by boat from Eriskay.

LINGAY (Sound of Harris Islands): *see under* Sound of Harris Islands.

LISMORE

SITUATION: in the lower waters of the sea-arm of Loch Linnhe, 6½ miles NW. of Oban.
AREA: 10 miles by 1 mile.
POPULATION: 350.
ACCESS: though Oban is the nearest large port, the mainland is only 200 yds distant and boats go to the island from Port Appin.

This green, sheltered isle is filled with wild flowers in summer: pink valerian, eye blue harebell, and fuchsia. Indeed the island's name is said to be derived from the Gaelic for 'great garden' or 'great enclosure'. At no point does it rise higher than 417 ft, and one of its curiosities is the raised beach which encircles the island. Dalradian limestone peeps out of the green, and the coastline has many caves, eaten out by the sea.

Lismore was once the seat of the diocese of Argyll and possessed a cathedral, a modest building which has long since been superseded by the parish church. The island was occupied in the 6th century by St Moluag, a Pictish Christian. Under the turf by the duns are the bones of Pictish kings and Vikings. Cairns mark the former Viking stronghold of Castle Coeffin.

A literary controversy that has never completely been stilled was touched off by the discovery of *The Book of the Dean of Lismore*, a 16th-century collection of Gaelic and English manuscript poems. These, it was argued, lent probability to the claims of James Macpherson and his supporters on the subject of the bard Ossian. Macpherson claimed that his two epics, *Fingal* (1762) and *Temora* (1763), were translations from Ossian. Dr Johnson attacked the authenticity of this claim with his customary violence, while Dr Blair supported the contention of Macpherson. The Highland Society of Edinburgh appointed a committee to inquire into the subject in 1797, but they arrived at no firm conclusion. Nevertheless there are those alive today who insist that the Dean of Lismore supplied a valuable link with an ancient culture.

Neat crofters' houses are dotted across the island, but the population is dwindling. On the W. coast, situated on a high rock, are the ruins of the episcopal castle of Auchindown, and at the S. extremity is a lighthouse. At the N. end is the ferry to Port Appin, which plies from a good harbour, Port Ramsay. Boats from Oban arrive at the S. end of Lismore.

Accommodation: various farms and crofters' cottages take in visitors.

LITTLE COLONSAY

SITUATION: 1 mile S. of Ulva in the Inner Hebrides.
AREA: ¾ mile by ½ mile.
POPULATION: uninhabited.
ACCESS: by boat from Mull, by special arrangement.

At one time Little Colonsay provided a living for several families derived from crops and grazing, but the brutal evictions general in the mid-19th century resulted in its loss of population. At this time crofters' grazing lands were forcibly turned over to sheep grazing, and though high profits were made for a while from this, the sheep were allowed to over-graze the land and much damage was caused. Tenants were not only turned off the island, but in some cases their homes were burned to the ground. But basically this is a fertile little island, and could still support a few people.

LITTLE CUMBRAE: *see under* Cumbraes, the.

LITTLE ROE

SITUATION: in Yell Sound, between Yell and Shetland Mainland.
AREA: 71 acres.
POPULATION: uninhabited.
ACCESS: by boat from Shetland Mainland.

The island, one of the Shetland Isles (q.v.), has some grazing land, and was inhabited and cultivated 200 years ago.

LITTLE ROSS

SITUATION: at the entrance to Kirkcudbright Bay, ¼ mile W. of the peninsula of Ross.
AREA: 2½ furlongs by 1 furlong.
POPULATION: uninhabited.
ACCESS: by boat from Ross, but permission must be obtained from the Commissioners of Northern Lighthouses, 28 North Bridge, Edinburgh, EH1 1QG, who own the island.

The lighthouse here was built in 1843. There are two houses on the island which have not been inhabited for many years.

LONGAY

SITUATION: 2 miles E. of Scalpay and 5½ miles W. of Erbusaig on the Kyle of Lochalsh, Inverness-shire.
AREA: ¾ mile by ½ mile.
POPULATION: uninhabited.
ACCESS: by boat from Scalpay.

LUCHRUBAN

SITUATION: close inshore to the Butt of Lewis, to which at low tide it is connected.
AREA: 80 ft by 70 ft.
POPULATION: uninhabited.
ACCESS: see Lewis.

This must surely be the smallest island to be listed in this book, but it is steeped in far more history than many larger islands. Sometimes called the Pigmies' Isle, Luchruban has a rounded, plateau-like surface, covered with thrift and short sea-grass. Dean Donald Monro, High Dean of the Isles, recorded in the 16th century that 'at the north poynt of Lewis ther is ane litle ile, callit Pigmies Ile, with ane litle kirk in it of ther awn handey wark. Within this kirk the anchients of that Countrey of Lewis says, that the saide pigmies has been eirded [interred] thair.'

The Dean claimed to have read accounts of these 'pigmies' in other ancient records, though he did not give chapter and verse for this. Later historians and archaeologists have, however, accepted his findings, and Buchanan in his *History of Scotland* referred to Luchruban as a place where 'a diminutive race have been buried, and many strangers, on digging deep into the earth, have found, and still find, small and round skulls, and little bones belonging to different parts of the human body.'

A sceptical Englishman, Captain John Dymes, made a survey of Lewis in 1630 but, though casting doubts on the legend of the pigmies, admitted that on the island he saw the walls of a chapel about 8 ft by 6 ft. He dug and discovered some bones beneath the floor of the chapel, but doubted whether such small bones could possibly belong to human beings. In Dymes's rough sketch-map of the island in the British Museum it is called 'The Ile of Pigmies'. An 18th-century map names it Ylen Dunibeg, a corruption of the Gaelic for 'Island of the Little Men'.

The present name, Luchruban, is thought to be a corruption of the Irish Lupracháan, meaning 'leprechaun'.

W. C. MacKenzie carried out extensive research into the legend of the pigmies early in the 20th century. His excavations on the island revealed the remains of two unroofed chambers, one circular and one oblong, connected by a passage. He was convinced that this was proof of the existence of the chapel. His other finds were less impressive, including some fragments of unglazed pottery (now in the National Museum of Antiquities in Edinburgh) and some bones – but the latter were declared to be those of oxen, sheep, a dog, and birds.

The mystery of the pigmies remains unsolved. Legend in Lewis says they were a race of diminutive people of Spanish or Iberian origin who arrived in about 500 BC.

Books: *The Book of the Lews*, by W. C. MacKenzie, 1919; Captain Dymes's account of his survey of Luchruban is to be found in the *Proceedings of the Society of Antiquaries*, 1904–5; *The Enchanted Isles*, by Alasdair Alpin MacGregor, Michael Joseph, 1967.

LUING

SITUATION: separated from Seil Island ($\frac{1}{4}$ mile to N.) by Cuan Sound, and 2 miles W. the Argyll mainland, in the district of Lorne.

AREA: $6\frac{1}{2}$ miles by $1\frac{1}{2}$–2 miles.

POPULATION: 338.

ACCESS: by the Cuan Ferry from Seil.

There is a road from N. to S. down the whole length of Luing, which has a small settlement on its SE. side. It is famed for its production of slate, and it was from Luing that the slates were obtained to re-roof Iona Cathedral. Today, however, only one man works at the craft of slate-cutting, and agriculture is the chief occupation. Luing cattle have in recent years been specially bred on the island.

The name Luing is pronounced without the 'u'.

Accommodation: available but limited.

LUNGA

SITUATION: the largest of the Treshnish Isles in the Inner Hebrides, 5 miles W. of Mull.
AREA: 172 acres.
POPULATION: uninhabited.
ACCESS: by boat from Mull or Iona.

Lunga was at one time used extensively for sheep grazing, but its inhabitants left the island 100 years ago. They left behind them swarms of rabbits and mice which still exist there.

The approach to Lunga is navigationally somewhat hazardous, owing to shallow water and hidden rocks. The best approach is through the channel between Lunga and Creag a'Chasteal Island and on to the shingle beach. It is possible (after obtaining permission) to camp on the high land where there are some sheltered hollows. The N. end of Lunga is the highest ground, rising to 337 ft, while the southern half is flat-topped, with a broad expanse of green. The cliffs average about 70 ft in height, but the best viewpoint for seeing the scattered mass of the Treshnish Isles (q.v.) is from the top of what is known as Lunga Hill.

Remains of old houses can still be seen on Lunga, notably on the N. terraces of the island, built there probably because of the proximity of fresh water. But this water supply is not always maintained when the weather is dry.

Sanctuary Rock is the haunt of a variety of sea birds, and is riddled with puffin holes. Fulmar also breed on Lunga, and the shag, eider duck, shelduck, oyster-catcher, ringed plover, kittiwake, razorbill, black-backed gull, guillemot, raven, wren, starling, rock pipit, and wheatear are all to be found. Surprisingly for so rocky an isle, flowers grow in remarkable profusion and variety, ranging from the primroses and violets in the ravines to the buttercups and daisies in the open, clumps of blue-bells, wild roses, honeysuckle, campion, sea pinks, and tiny flowers of an Alpine type in the cliff crevices. There is also a comparatively rare plant with a blue flower known as oyster plant.

For the adventurous there is a tunnel in the rocks to explore, leading to a cavern filled with shells. Half-way across the island its waist narrows and dips so that it is only a few feet above sea level. Fishing for lobsters and crabs is excellent.

For permission to land on Lunga application should be made to the owner, Lady Jean Rankin, House of Treshnish, Calgary, Isle of Mull.

MAMHA CLEIT: *see under* Sulasgeir.

MAY

SITUATION: off the coast of Fife, 6½ miles E. of Anstruther.
AREA: 126 acres.
POPULATION: 14.
ACCESS: by boat from Pittenweem on the Fife coast.

May is one of the chief centres in the E. of Scotland for recording the autumn and spring migrations of birds. There is a bird-watching station, and also a lighthouse, on the island. Bird-watchers from all over the United Kingdom visit May, which is a favourite nesting-place for birds travelling both N. and S. It is scheduled as a nature reserve and its fertile soil provides good grazing.

Like other Scottish islands, May attracted the attention of the early Christian missionaries, and St Adrian was murdered here in the 9th century by a party of Danish raiders. There are the ruins of a 13th-century chapel which is dedicated to this saint.

Accommodation: this is available for small parties of bird-watchers and others.

MEALASTER

SITUATION: 1½ miles W. of Lewis in the Outer Hebrides, 4½ miles SSE. of Brenish.
AREA: 1 mile by ¾ mile.
POPULATION: uninhabited.
ACCESS: by boat from Lewis, but landings are not always easy.

MEALL BEAG: *see under* Badcall.

MEALL MOR: *see under* Badcall.

MINGULAY

SITUATION: in the Outer Hebrides, 1½ miles S. of Pabbay and ¾ mile N. of Berneray.
AREA: 2½ miles by 1 mile.
POPULATION: 4.
ACCESS: by boat from Barra by special arrangement.

Mingulay is not easily accessible except in good weather. It is a bleak isle with cliffs rising to 800 ft, and these are fissured by whin dikes so that boats can pass between their walls. On the W. side of Mingulay is a natural rock bridge 550 ft above sea level.

There is little vegetation, but the isle is worth a visit if only to inspect its bird life, notably the black-backed gull (which takes six years to mature) and the puffin. For centuries Mingulay was inhabited by crofters who eked out a bare living and subsisted mainly by eating sea birds and their eggs.

Poverty and desperation caused many of them to emigrate some sixty-five years ago to nearby Vatersay. They were known as the 'Vatersay Raiders', for their trek to that island was in the nature of an invasion. They built themselves wooden cabins and refused to leave, despite threats from the owner, Lady Cathcart Gordon. In 1908 they were brought to court and sentenced to six months' imprisonment in Calton Jail, Edinburgh. But they only served six weeks of this period.

See also Berneray.

MONACH ISLES

SITUATION: 8 miles SW. of Houghharry Point, North Uist, in the Outer Hebrides.

TOTAL AREA: 600 acres.

POPULATION: occasional only, mainly lobster fishermen and crofters from Lewis.

ACCESS: by boat from Tighary, North Uist.

These tiny, low-lying islands total six, three of which – An Ceann Iar, Crois Shitinis, and An Ceann Ear – are connected at low tide by sandy beaches. The other (and smaller) islands are Sillay, Srogay, and Sgrot-Mhor.

Today they are used for sheep grazing and lobster-fishing, but they have a history of habitation. None of the islands rises to more than 50 ft above sea level, but all have a remarkably springy turf and a wide variety of flowers, such as bird's foot trefoil, thyme, heartsease and kidney-vetch. Marram grass stabilises the sand dunes. Some Arctic terns nest in the sand.

As long ago as 1263 the islands supported a fair-sized population. A nunnery was established on An Ceann Ear, attached to the ecclesiastical outposts at Iona (q.v.). The nuns had a reputation for being Amazonian in strength and skill, capable of handling large boats which they rowed to North Uist, bringing back loads of peat for fuel. A monastery was also set up at Sillay, the most westerly island of the group, where the monks maintained a permanent light to guide passing boats.

The Reformation saw an end to the religious institutions on the islands, which passed into the hands of the family of the Lord of the Isles, then to Lord James MacDonald of Sleat in Skye, and in the 19th century to Major Iain P. Orde. The last-named passed his property on to the Duke of Hamilton in 1944, and the present owner is Lord Granville.

In the late 16th century the Monach Isles were said to have a population of about 100. By all accounts the islands were then sufficiently fertile to possess 1,000 cattle, as well as many sheep. Early in the 19th century, due to soil erosion, the islands lost their fertility and almost all the people left the Monachs. It was after this that marram grass was planted in an effort to check further erosion, and as a result by 1891 the population had risen to 135, including twelve lighthouse keepers and their families.

At the beginning of the present century the population again began to fall steadily, until in 1942 only two families were left, and they were evacuated the same year. A lighthouse 135 ft high was erected on Sillay in 1864, but this was closed during World War II.

As in the Scillies, the inhabitants for many years looked to wrecks and wreckage washed ashore to supplement their income. Records suggest that such wrecks as that of the *Inflexible* (in 1894) and the *Vanstable* (in 1903) supplied the islanders with many goods of which they were sorely in need. Certainly the Monach Isles provided themselves with amenities out of all proportion to their size – a post office, a school, and a mission church. Today the lobster fishermen from Lewis on their visits to the island occupy the cottages evacuated by the earlier inhabitants.

MOUSA

SITUATION: 1 mile E. of Sandwick on Shetland Mainland.
AREA: 2 miles by ¾ mile.
POPULATION: uninhabited.
ACCESS: by boat from Sandwick.

A small, undistinguished island in the Shetland Isles (q.v.), significant solely because it possesses the most perfect specimen of the Pictish *broch*. This is known as Mousa Castle, and is 45 ft high, tapering rather like a miniature lighthouse from a base of 158 ft in circumference.

Mousa Castle is not only well-preserved, but quite unique. The centre is an open shaft letting in light and air to the galleries. It is different from other similar *brochs* and towers in that it decreases in width up to about 10 ft from its top and then expands outwards. The object of this was to make it impossible for enemies to scale the summit, while the small entrance door made an onslaught from that quarter extremely difficult. It is about 1,800 years old.

Legend has it that Erland, son of Harald the Fair-Spoken, carried off to Mousa Castle in about 1150 the mother of Harald, the Norwegian jarl – a famous beauty. The jarl, being unable to take the castle by force, finally agreed that his mother should become the wife of her captor.

There is, however, another legend about Mousa in *Egil's Saga* – one of the great Icelandic sagas. It tells how a young man named Bjorn, who had seduced a girl named Thora in Norway, went ashore on Mousa about AD 900, and how he and his girl-friend made their home in the *broch.*

Mousa is now inhabited only by a flock of sheep and a herd of ponies which run loose. It was farmed by a resident farmer in the last century, and is now owned by Robert Bruce, Lord Lieutenant of Shetland.

Book: *Royal Commission on the Ancient Monuments and Constructions of Scotland, Inventories and Reports, Orkney and Shetland*, 3 vols, revised edition, 1946.

MUCK

SITUATION: 6 miles NNW. of Ardnamurchan, Inverness-shire, in the Small Isles group of the Inner Hebrides.
AREA: $2\frac{1}{2}$ sq. miles.
POPULATION: 15 (1966).
ACCESS: no steamer calls at Muck, and there are only occasional motor-boat connections with the nearby island of Eigg, 3 miles to the N.

This is the smallest island of the group known as the Small Isles (q.v.), and it is also the most isolated. The name 'Muck' derives from the Celtic word for pig, and in fact the island was called 'Swyne's Isle' by Dean Monro in his book *Western Isles of Scotland.*

There is a small harbour at Port Mor on Muck, but the traffic is limited. The island itself is verdant and pleasing, rising to a height of 451 ft, and has some sandy beaches. Excellent lobster fishing is to be found here.

Muck has owed much to its enlightened laird, Laurence

MacEwen, who planned its farming successfully over a lengthy period and enabled its population to be maintained. There is no accommodation for overnight visitors.

MUCKLE FLUGGA

SITUATION: ½ mile N. of Unst in the Shetland Isles.
AREA: 12 acres.
POPULATION: 2.
ACCESS: by boat from Unst.

The most northerly island in the British Isles, little more than a remote rock in the Shetland Isles (q.v.) on which is situated a lighthouse. It can thus claim to be the most northerly inhabited island of the United Kingdom.

MUCKLE ROE

SITUATION: close inshore to Shetland Mainland, to which it is now linked by a narrow bridge across Roe Sound.
AREA: 7 sq. miles.
POPULATION: 103.
ACCESS: by bridge from Shetland Mainland.

This island, which is almost circular in shape, is remarkable for its particularly high, red cliffs. Its centre is marked by a hill, South Ward, which rises to a height of 557 ft. It is well worth a visit if only for the magnificent coastal scenery, especially the Hams in the NW.

This is a farming community, and most of the inhabitants live close to Roe Sound or on the E. side of Muckle Roe.

See also Shetland Isles.

MUGDRUM

SITUATION: in the Firth of Tay, ¼ mile N. of the coast of Fife, 2 miles NE. of Abernethy.
AREA: 35 acres.
POPULATION: 4.
ACCESS: by boat from Mugdrum village on the Scottish mainland.

Mugdrum belongs to the parish of Newburgh and lies opposite the village of Mugdrum on the Fife coast. It contains arable land of the richest quality. It was at one time extensively farmed, and had a population of twenty-six in 1891.

MULDOANICH: *see under* Barra Isles.

MULL

SITUATION: 8 miles W. of Oban, in the Inner Hebrides.
AREA: 353 sq. miles.
POPULATION: 3,389 (1966).
ACCESS: in summer there is an air service from Glasgow and Oban to Glenforsa (Loganair Ltd); a car-ferry service runs from Oban to Craignure, and a steamer from Oban to Tobermory.

The third largest of the Hebridean islands, Mull is 30 miles long, but its indented coastline extends for 300 miles. Its scenery varies from verdant valleys to high mountain peaks in the S. and E. The highest mountain is Ben More (3,169 ft). There are extensive moorlands and some excellent beaches on the W. coast.

Mull: Tobermory (Radio Times Hulton Picture Library)

The Sound of Mull, which separates the island from the mainland, is a sheltered stretch of water. Buses run to all parts of Mull. Great efforts have been made in recent years to bring tourism to the island, but there has been some considerable opposition to this on the grounds

opposite: *Mull: Ben More from across Loch Na Keal (British Tourist Authority)*

that money could be more wisely spent in the long run on developing farming and forestry projects. The capital of the island is Tobermory (population 617), which once had 2,000 inhabitants. Its importance as a shipping centre has dwindled owing to the development of a modern pier at Craignure.

Tobermory (its name means 'The Well of Mary') has a sheltered bay and anchorage, and its houses rise in attractive terraces above the little bay. It was off here that one of the ships of the Spanish Armada was lost in 1588: there have been a number of unsuccessful attempts to salvage her. The town was originally founded as a fishing village under the auspices of the British Society for Encouraging Fisheries, but there was too much competition for this scheme ever to develop economically. Yet there is varied fishing around Mull – herring, cod, mackerel, saithe, ling, rockfish, flounders, plaice, turbot, sole, and skate are all caught off these coasts.

Mull is a splendid island for the rambler who likes mountains, wide expanses of moorland, and pastoral beauty. But it lacks the surprises of Skye and other Hebridean isles, and has curiously failed to draw favourable comment from writers and artists: Dr Johnson found it 'a dreary . . . a most dolorous country', which is a somewhat unfair comment. Its inhabitants outside of Tobermory live in isolated sheep farms and forestry centres. In the SW. is the village of Fionphort, and a few miles inland lies Bunessan, but the other settlements are small and insignificant except for Dervaig, in one of the least spoilt areas of the island. Lochdonhead is only a group of houses around a tiny bay; Craignure is merely the terminus for the car ferry from Oban; and Salen is a straggling village.

The most attractive views are to be had from Torloisk, around Loch Na Keal and Loch Tuath. There are streams everywhere, and waterfalls such as Eas Forss at the E. end of Loch Tuath. Unusual features of the sea lochs of Mull are the miniature waterspouts which are whipped up in gales.

There is much on the island to interest the geologist. An outcrop of lignite, a kind of coal, in a three-foot seam at Beinn An Aoinidh has been worked, but only for domestic and local use. Then there are the 'leaf beds' at Ardtun, near Bunessan – an accumulation of fossil leaves and petals in a layer of mudstone lying between two early lava flows. Many caves exist along the coast, notably Nun's Cave at Carsaig, Lord Lovat's Cave near Loch Buie, and the Carsaig Arches W. of Carsaig. Mull granite, pink and red in colour, is also notable, being found in the rock and capable of taking a high polish.

Relics of the Stone Age have been found on Mull. Several sites of mesolithic and Bronze Age occupation have been examined and recorded, including cairns at Loch Buie and Dail Na Carriagh, and a stone circle at Duart. Particularly interesting is the Druid's Field below Craig-a-Chaisteal. There are also remains of Celtic forts, notably that at Dun Nan Gall on the N. shore of Loch Tuath, about 3 miles W. of Ulva ferry, and there are pagan stones on which Christian crosses have later been carved to mark the pilgrims' way to Iona (q.v.).

Undoubtedly Mull also suffered an invasion by the Norsemen, though there is little hard evidence of this. It is also possible that Norman influence reached Mull, as some island families can claim Norman descent. One of the oldest monuments of the island is Duart Castle, the home of the chieftain of Clan Maclean. He was dispossessed of it in 1692 and the castle became a ruin, but was restored in 1912 by the chief of the clan, Sir Fitzroy Maclean. It dates back to the 13th century and commands the Sound of Mull. The walls of its tower are 10–14 ft in thickness.

Other castles worth visiting are Aros Castle at Salen Bay, and Moy Castle, the former abode of the MacLaines of Lochbuie. The latter castle was occupied until the middle of the 18th century, but is now barred to visitors on account of the risk of falling masonry.

The Macleans were the principal clan in the Mull area, but their support for the Jacobite cause led to their being dispossessed of their lands, which were handed over to the Duke of Argyll, in the 18th century. But by the middle of the following century the Dukes of Argyll had been forced to part with their lands on Mull owing to failure to encourage industry in the island. The island's population rose to 10,612 by 1821, but a decline set in shortly after this date. Crofters' lands were confiscated and turned over to rented sheep grazing, and many tenants were evicted and their houses burned. Destitute families migrated to Tobermory, where a poorhouse was built to cope with them. By 1871 the island's population had fallen by half, and many families had emigrated to America, Canada, or Australia. The Crofters' Acts of 1886 and 1892 did slightly ameliorate conditions for the people, but in the long term little was achieved because nothing was done to enable the crofters to acquire sufficient land to be economically viable.

Mull is a favoured centre for yachting, and some excellent sailing can be obtained in the vicinity. There are also facilities for golf, pony-trekking, skin-diving and, to a limited extent, climbing.

Accommodation: hotels, boarding houses, etc., provide accommodation for about 600 visitors at a time. There is a 60-bedroom hotel at Craignure, sponsored by the Highland Development Board and run by the Scottish Highland Hotels group. There is also varied accommodation at Tobermory. Inquiries should be made at the Tourist Association and Information Centre in Tobermory.

Post Offices and banks: available at Tobermory.

Books: *The Isle of Mull*, by P. A. MacNab, David & Charles, 1970; *The Beautiful Isle of Mull*, by Thomas Hannan, 1926; 'On Tertiary leaf-beds in the Isle of Mull', by the Duke of Argyll, *Quarterly Journal Geological Society*, vol. vii (1851); *The Island of Mull: its History, Scenes and Legends*, by John MacCormick, 1923.

MULLAGRACH

SITUATION: one of the Summer Islands, 2 miles W. of the Ross and Cromarty mainland.
AREA: 185 acres.
POPULATION: uninhabited.
ACCESS: by boat from Achiltibuie.

MURRAY'S ISLE: *see under* Isles of Fleet.

NIBON

SITUATION: close inshore and SW. of Northmaven, Shetland Mainland.
AREA: 62 acres.
POPULATION: uninhabited.
ACCESS: by boat from Shetland Mainland.

NORTH HAVRA

SITUATION: 3 miles W. of Scalloway, Shetland Mainland.
AREA: 50 acres.
POPULATION: uninhabited.
ACCESS: by boat from Scalloway.

NORTH RONA

SITUATION: 44 miles NNE. of the Butt of Lewis in the Outer Hebrides, and 45 miles NW. of Cape Wrath in Sutherland.
AREA: 300 acres.
POPULATION: uninhabited.
ACCESS: by boat from Lewis when weather permits.

This is the most northerly island of the Outer Hebrides, dominated in the centre by the high ridge known as Toa Rona (355 ft). One of its chief attractions is its many caves, one of which, Tunnel Cave, has a blow-hole 100 ft long. The approaches to the island on the N., SW., and SE. are strewn with reefs and rocks.

The island takes its name from St Ronan, who is said to have gone to the island to escape from the presence of chattering women on Lewis! However, as there are many saints bearing the name Ronan, it is difficult to know whether or not this particular legend has any foundation. The island was inhabited from about the 8th century, probably by roving Norsemen in the first place. North Rona seems to have been able to support a modest population in its earlier history, for in 1549 Sir Donald Monro wrote: 'this ile is half mile lang, half mile braid: abundant of corn growis in it be delving, be abundante of naturall claver girs for scheip.' He went on to describe the chapel of St Ronan, the remains of which can be seen there today.

Certainly North Rona was more productive than its neighbouring island, Sulasgeir (q.v.). In Monro's time there were sheep and cows on the island, with a bull for breeding, and crops of oats, barley, corn, and potatoes, while seals and wild fowl as well as fish supplemented the diet. The population then normally comprised about five or six families, and rarely exceeded thirty. Towards the end of the 17th century the island suffered from a plague of rats which ate all the crops and reduced the inhabitants to poverty. The final blow came when a marauding ship landed and took away their one bull.

North Rona never recovered from these disasters, and in 1797 it was recorded that there was only one family living on the island. In 1844, when Sir James Matheson purchased the island of Lewis, he offered North Rona to the British Government as a penal settlement. The Government declined the offer and North Rona remained desolate for many years. Then in 1884 two men from Ness settled on the island. A year later they were found dead by a visiting party. A memorial stone marks their burial site and bears the inscription:

> Sacred to the memory of Malcolm M'C Donald Ness who died at Rona Feb 18 1885 aged 67 also M M'C Kay who died at Rona same time. Blessed are the dead who die in the Lord.

Visitors have formed strangely contrasting opinions about North Rona. The Duchess of Bedford, who visited it in 1910, complained of 'the all-pervading stench of the nesting-places of hundreds of fulmars, great and lesser black-backed gulls and herring gulls'. On the other hand Malcolm Stewart, author of a book on the island, gave a much more generous impression:

> No one could ever fail to experience immense pleasure on first landing on North Rona. A narrow strip of land, on the north end covered with green vegetation and surrounded on three sides by the ever-present Atlantic. . . . Ruins of villages deserted for many a year usually induce a feeling of despondency and regret, not to mention solitude, but here . . . no one has such thoughts of sadness.

The roofs of the ruined village have tumbled down and grassy turf has covered what remains. St Ronan's chapel or cell is of very ancient construction which bears traces of the megalithic influence.

Books: *Ronay*, by Malcolm Stewart, Oxford, 1933; *St Kilda and other Hebridean Outliers*, by Francis Thompson, David & Charles, 1970; *A Description of the Western Isles of Scotland called Hybrides*, by Sir Donald Monro, 1549, reissued 1933.

NORTH RONALDSAY

SITUATION: the most northerly of the Orkney Islands, 2½ miles N. of Sanday.
AREA: 3½ sq. miles.
POPULATION: 120.
ACCESS: by the Orkney Islands Company ship from Kirkwall (weekly service), or by air by the Loganair (twice weekly) service connecting with Glasgow, Edinburgh, and Aberdeen. There is also a small-boat link with Sanday.

This is the most remote of the inhabited Orkney Islands (q.v.) and, bearing in mind its small population, it has somewhat better amenities than many of the other islands – a school, a resident doctor, mains water, two shops, and several telephone links.

North Ronaldsay is a very low-lying island and is exposed to the full force of the Atlantic gales. It bears marks of the presence of prehistoric man, including a number of Bronze Age burial cists; and in the SE., on the shore of Strom Ness, the Iron Age *broch* of Burrian, an impressive structure, still stands, though partly destroyed by the sea. Its wall is 15 ft high on the landward side, and when excavated in 1870 it also revealed traces of Celtic Christian associations, including the Burrian Cross, inscribed on a flat piece of stone.

Nearly all place-names here are of Norse origin, and the ancient name of the island was Rinansay, as it is frequently called in *The Orkneyingers' Saga*. Crofting is still the principal occupation, there being some 2,000 native Orkney sheep on the island. It is an estate island, farmed by tenants, with three farms and sixty-seven crofts. In 1831 the population rose to a peak of 522, due almost entirely to the estate owners encouraging the production of kelp. Later there were large-scale evictions of tenants which caused much distress, though improving farming standards.

Today, apart from sheep farming, there are some 300–

400 cattle on the island. The water supply is drawn from the Loch of Anchum. There is no mains electricity, but good roads have been provided by the estate owners. The lighthouse is 184 ft high – the second tallest in Scotland – and is the outstanding landmark.

NORTH UIST

SITUATION: in the Outer Hebrides, 8 miles SW. of Harris and linked to Benbecula by a causeway.

AREA: 118 sq. miles.

POPULATION: 2,579.

ACCESS: by air from Glasgow, Aberdeen, or Inverness to Benbecula; or by sea from Uig in Skye to Lochmaddy.

In contrast to South Uist, all the inhabitants of North Uist are Protestant, and they are much more akin to the predominantly Norse peoples of Harris and Lewis to the N. than to those of their southern neighbour. The island is not as mountainous as South Uist, though it has three hills, of which Eaval (1,138 ft) is the highest. But like both Benbecula and South Uist, it is covered with inland lochs, especially on the W. side.

Lochmaddy is the principal port and village, and the last port of call for ships leaving for lonely St Kilda. The loch itself, though only 5 miles long with an entrance a mile wide, has such a twisting and tortuous coastline that mathematical experts assert that its true length is 50 miles. The loch itself contains several tiny islets.

Fishing and angling on the island are particularly good. The hotel at Lochmaddy has the fishing rights for most of the island and every angler who steps out must, by custom and regulation, present his second fish to the hotel, keeping the rest of the catch for himself. The Department of Agriculture and Fisheries also has some trout lochs in North Uist where fishing costs about 50p per day. The salmon fishing here is the best in the Hebrides.

The eastern part of the island has high hills and moorland, interspersed with many lochs, while on the W. side there are long sandy beaches, ideal for swimming and surfing. Surprisingly, too, the relatively mild climate enables tulips and daffodils to be grown, and North Uist boasts a flourishing bulb-growing industry.

Many remains of ancient settlements are to be seen in North Uist, particularly the chambered Cairn of Barpa Langass, a burial construction 18 ft high and 72 ft across, entered by means of a tunnel. It is situated near the S. side of the main road some 6 miles W. of Lochmaddy. Almost equally interesting are the ruins of a church called Teampull Na Trionaid, built about the 14th century with a roof part vaulted and part thatched, lying on a rise close by Carinish.

The harbour at Lochmaddy is a particularly good one, having a modern pier and providing ample space and shelter for a large number of craft. The Sheriff Court here deals with all minor legal cases arising in the islands of North and South Uist and Benbecula.

Accommodation: there is a good hotel at Lochmaddy and 2 others on the island, as well as various boarding-houses and some cottages offering bed and breakfast. There are also some cottages to let furnished in the summer months. For further information write to the Tourist Office, Lochmaddy, North Uist.

Book: *The Scottish Islands*, by G. Scott-Moncrieff, Oliver & Boyd, 1965.

NOSS

SITUATION: ¼ mile E. of Bressay in the Shetlands.
AREA: 1¼ sq. miles.
POPULATION: uninhabited.
ACCESS: by boat from Lerwick, Shetland Mainland.

Perhaps the most popular bird sanctuary in the Shetlands (q.v.), Noss attracts hundreds of visitors every year to see its huge population of storm petrels, guillemots, gannets, puffins, shags, razorbills, and a wide variety of terns.

There is some magnificent cliff scenery on Noss – on the E. side of the island the Noup rises precipitously to a height of 592 ft. There is also the Holm, a stack of a rock only 160 ft high, but too steep to climb, though it is recorded that in the 17th century it was climbed for the wager of a cow. The daring man who made this successful attempt fixed up a rope-railway strong enough to carry a wooden cradle to the summit of the Holm, which is flat and covered with turf. Alas, he fell to his death while climbing down, but the rope-railway was used until the end of the last century.

In 1841 there were twenty-four inhabitants on Noss, but today the only traces of habitation are the ruins of former cottages. There is, however, one shepherd's cottage that is habitable still. One word of warning: the great skuas, or bonxies as they are called in Shetland, can be something of a menace in nesting time, and are known to make ferocious attacks on humans. It is advisable to be armed with a stick or umbrella to ward off their sudden swoops. They will sometimes attack in dozens. The skuas also take heavy toll of the kittiwakes and guillemots.

Gannets begin to nest on Noss in March, and hatching usually occurs from May onwards: they disperse over a wide area of the Atlantic in October.

Book: *Islands by the Shore*, by Alasdair Alpin MacGregor, Michael Joseph, 1971.

OIGH-SGEIR

SITUATION: 10 miles W. of Rhum in the Inner Hebrides.
AREA: 7 acres.
POPULATION: uninhabited.
ACCESS: by boat from Rhum.

A scattering of tiny rocky islets comprise the low-lying island of Oigh-sgeir, pin-pointed from a distance solely by reason of its tall, slim, white lighthouse. This lighthouse, which is 128 ft high, has a group-flashing light of three quick flashes every thirty seconds, and is visible almost 20 miles away.

Oigh-sgeir never rises to a greater height than 34 ft above sea level, and is composed of pitchstone porphyry, covered by a thin layer of grass and thrift. It is said that at one time lighthouse men kept goats on the isle.

OLDANY

SITUATION: in Eddrachillis Bay, close inshore and 3 miles E. of Point of Stoer, Sutherland.
AREA: 2 miles by 1½ miles.
POPULATION: uninhabited.
ACCESS: by small boat only from the mainland.

After Handa (q.v.) this is the largest island among the many in Eddrachillis Bay, but the land is not fertile, though it can support a certain amount of grazing as it is covered with grass. There is a narrow passage of sea between Oldany and the Scottish mainland which can be navigated by a small boat, taking care to avoid the patches of rock in the vicinity. Oldany rises to a height of 329 ft.

See also Eddrachillis Islands.

ORKNEY ISLANDS

SITUATION: a group of islands, the nearest of which is 8 miles, and the furthest 50 miles, N. of John o' Groats in Caithness.
TOTAL AREA: 376 sq. miles.
POPULATION: 18,743.
ACCESS: by air (BEA) from Glasgow, Edinburgh, Aberdeen, or Inverness to Wick, Caithness; and by sea from Scrabster in Caithness to Stromness, Orkney Mainland. Car-ferry service. There are also regular inter-island steamer sailings and light aircraft services, with a daily air service to Kirkwall from Wick.

The Orkney Islands form an entire county. There are sixty-seven islands, but Orkney Mainland, the chief island, has a greater acreage than all the other islands combined. Only twenty-one of the islands are inhabited.

The Orkneys were first colonised by the Picts, and later by the Norsemen. In 1468 they were pledged by King Christian I of Norway as part of the dowry of his daughter Margaret when she married James III of Scotland. They are rich in Stone Age remains – notably the Stone Age village of Skara Brae and the chambered burial mound of Maeshowe.

Remote as they are, the Orkneys are easily accessible to the visitor on account of the regular steamer and air services which link them with most parts of the British Isles. They have the attraction in summertime of being in latitudes where daylight lasts so long that night itself rarely becomes more than a somewhat overcast day. True, a poet, Captain Hamish Blair, wrote of Orkney, 'All bloody clouds and bloody rains,' but although cool in summer, the islands can have brilliant sunshine.

There is nowhere better for sea-trout fishing in the British Isles, and some of the best brown-trout fishing is also to be had in Orkney. What is more important, as there are very few laird-landlords, permission to fish the lochs is usually easily obtainable at a modest cost. Facilities are provided for sea anglers.

Orkney speech has its roots in the Norse language, as its place-names and surnames suggest. There is some controversy about the origin of the name Orkney, but the general view is that it comes from the Celtic word *orc*, meaning 'wild boar', and the Norse *ey*, 'island'. The combination of the two makes it 'the islands of the wild boar people'. But this is perhaps too simple a solution to what may be a much more complex problem of nomenclature.

Orkney Norn is the name given to the language spoken here during the Norse occupation and for centuries afterwards – a relic of the language peculiar to the Norwegians who lived in the fjords of SW. Norway. The Orcadians of today speak English with a kind of sing-song accent: their ancient version of Norse disappeared some time during the 18th century, though a few phrases of it can still be heard on some islands.

As a people the Orcadians are totally different from those of the Outer Hebrides. They are less moody, more outward-looking, and more stolid. The majority belong to the Church of Scotland.

If one is going to Orkney by sea, the route by the passenger and car-ferry service from Scrabster in Caithness to Stromness across the Pentland Firth is the most impressive. This route takes one past the Old Man of Hoy, towering 450 ft above the sea. There is also an alternative route by sea from Aberdeen, but this takes nearly four times as long and is a less interesting voyage. Those travelling by road can leave their cars at Wick airport and go on by air to Gremsetter Airport at Kirkwall, the flight taking only a few minutes.

Orkney has produced a number of writers who have achieved literary fame. Two of the best-known are Eric Linklater (born 1899), the novelist, essayist, and playwright, and Edwin Muir, the poet (1887–1959). The latter gives a remarkably vivid picture of Orkney in his *Autobiography*. Joseph Storer Clouston, though a novelist, covered the Orcadian scene in considerable historical depth in his *Records of the Earldom of Orkney* (1914).

See also Auskerry, Burray, Cava, Copinsay, Eday, Egilsay (Orkneys), Eynhallow, Fara (Hoy), Fara (Westray), Flotta, Gairsay, Graemsay, Hoy, Hunda, North Ronaldsay, Orkney Mainland, Papa Stronsay, Papa Westray, Rousay, Rysa Little, Sanday, Shapinsay, South Ronaldsay, Stronsay, Sule Skerry, Switha, Swona, Westray, Wyre.

Accommodation: this is mainly to be had in Kirkwall, where there are hotels, but a number of other islands also cater for visitors.

Post Offices: in various centres.

Books: *An Orkney Tapestry*, by G. M. Brown, 1969; *The Orkney Parishes*, by J. Storer Clouston, 1927; *History of Orkney*, by J. Storer Clouston, 1932; *The Orkneyingers' Saga*, by G. W. Dasent, 1894; *Summers and Winters in the Orkneys*, by D. Gorrie, 1868; *Orkney and Shetland*, by Eric Linklater, Robert Hale, 1965; *Orkney*, by H. Marwick, 1951.

ORKNEY MAINLAND

SITUATION: the largest of the Orkney group of islands, 20 miles N. of Caithness.

AREA: 23 miles by 5 miles.

POPULATION: 4,500.

ACCESS: by sea from Scrabster, near Thurso, to Stromness (2½ hrs), or from Aberdeen to Kirkwall (11 hrs). Local air connections from Wick to Grimsetter (Kirkwall).

Formerly called Pomona, which name still appears on some maps, Orkney Mainland is the centre of Orkney's county administration, and its chief town, Kirkwall, is also the capital of the Orkney Islands (q.v.).

The eastern part of Mainland is joined to the western part by the isthmus of Scapa, at the N. end of which

Kirkwall is situated. The island is divided into fourteen parishes. The parishes differ greatly in size and in distribution of population. For example, in Sandwick the houses and farm buildings are evenly distributed over the whole area, with neat fields in between, whereas in Harray the houses are spread out in haphazard fashion. Harray is to some extent enclosed by low hills, and contains a number of lochs. But the whole terrain is one of dispersed farms and cottages; there is no concentration of homes into villages. Farming is the chief occupation, beef cattle being the backbone of the industry in these parts. Most cultivated land is devoted to providing food for the cattle – grass, or fodder crops such as oats and roots. The cattle are exported for sale in Aberdeen, though some are kept for dairy farming. Sheep are also to be found on many farms. Fishing is very much a secondary occupation, concentrated principally in Stromness, where fishermen have established their own co-operative for handling lobster and crab.

Most of the population is concentrated in Kirkwall, the oldest town on the island, dating back to the days of the Norse occupation. It has a surprisingly large number of shops for so small a town, and even a floating bank which serves those islands unable to support a permanent branch. Kirkwall now is much the same as it was when described by Murdoch Mackenzie in 1750 – it lies 'along an arm of the sea . . . and consists of one Street three quarters of a mile long . . . all the houses are built of stone, some of them tolerably handsome both within and without, and for the most part have a kitchen garden behind them.' This famous Street runs from the harbour to St Magnus's Cathedral and includes the main shopping centre.

In front of St Magnus's Cathedral is an open space called Kirk Green, once used as a market-place and for holding the annual fairs. The name 'cathedral' is really a misnomer today: it is strictly only the parish church of the Church of Scotland. Built by Earl Rognvald in 1137, it is the property of the burgh of Kirkwall and not of the

Orkney: Earl Patrick's Palace (J. Allan Cash)

Church of Scotland. This is because it was originally built on Earl Rognvald's private land, and when James III made himself Earl he gave the church to the burgh.

The stonework of the cathedral is a patchwork of reds, yellows, browns, and greys which, externally, is aesthetically pleasing in certain lights, but incongruous in others. Inside stone of the reddish colour predominates, and is most effective. Much restoration work was carried out in the 1920s and 1930s. The tower is 14th century, and the spire was built in the present century.

The remains of St Magnus, who was murdered on the isle of Egilsay in 1116, were brought to the cathedral and buried there, but they seem to have disappeared during the Reformation. Bones discovered in the cathedral in the 18th century and again in 1919 are believed to have been those of Earl Rognvald and St Magnus respectively.

Also of interest are the Bishop's Palace and Earl Patrick's Palace. These are to the S. of the cathedral. The Bishop's Palace is the older building, but was reconstructed in 1541–8. Much more interesting is the Renaissance-style Earl Patrick's Palace, created by Earl Patrick Stewart, a 16th-century tyrant who exploited Orkney for his own personal gain, and was arrested for treason in 1609. Nevertheless he built a palace that was architecturally elegant and one of the finest of its kind in all Scotland. Here he entertained in great style with minstrels, trumpeters, and a magnificently-dressed body-guard. As Daniel Gorrie wrote, the palace 'possessed . . . all the features of a robber's stronghold though adorned with the elegances of a palace.'

One ancient feature of Kirkwall life which is still maintained is the 'Ba' Game' which takes place on New Year's Day. It is a football match played between those who live at the harbour end of the town and those at the Laverock end, the 'Uppies' and 'Doonies', as they designate themselves. Rather like Shrove Tuesday foot-ball in parts of England, it is a game in which as many as 150 people take part at a time. It can last for as long as five hours.

There are two hospitals in Kirkwall and two old people's homes, while eight resident doctors practise on the island, a high number for a small population. Facilities for entertainment are limited, there being only one cinema and the Orkney Arts Theatre, which was opened in 1968. Community life is typified by such organisations as the Orkney Field Club, several Young Farmers' Clubs, and the Orkney Heritage Society, which is concerned with preservation of ancient monuments and relics of the island's past. Sporting activities include football, rugby, golf (one club at Kirkwall and one at Stromness), squash, badminton, darts, bowling, and motor-cycling (several clubs).

opposite: *Orkney: Kitchener Memorial on Marwick Head (J. Allan Cash)*

Kirkwall has its own weekly newspaper, the *Orcadian* (founded 1854), which sells something like 9,000 copies a week, many of which are posted to Orcadians living overseas.

Stromness, the island's other principal town, is primarily a port, yet curiously much more attractive in some respects than modern Kirkwall. The town itself only came into existence in the early part of the 18th century when its port began to develop. After the Act of Union was passed Stromness developed a trade of its own and became a primary port for trade between Scotland and the Baltic. It also achieved importance as the last port of call for ships of the Hudson Bay Company and for whalers heading for the Davis Straits. One of the reasons for its importance was that the northern route to Britain was favoured by shipping because of its freedom from the pirates who threatened the English Channel and even the Irish Sea until late in the 18th century. Today Stromness has a population of about 1,700 and its importance has dwindled, for once it had a larger number of inhabitants than Kirkwall.

Nobody visiting Orkney should miss the excavated prehistoric village of Skara Brae, situated 6½ miles from Stromness on the W. of Orkney Mainland. It is as interesting to the visitor without archaeological knowledge as to the specialist. Skara Brae is at the S. end of the Bay of Skaill in Sandwick. It must have been buried in a sudden fierce sand storm for, when excavated, there was evidence that some of the inhabitants of this primitive settlement had tried to escape after being trapped, and also indications that those who did escape returned later and lived in the half-buried site. The remains of fires and the bones of animals they had cooked were found at three levels above the buried hut floors.

Today Skara Brae consists of the remains of ten well-preserved huts, bereft of their roofs: they are of rectangular shape with low, covered passages connecting them. The building is of unmortared flagstone. What is of most interest is the primitive furniture inside them –

Orkney: Skara Brae, 'the most perfect Stone Age village in Europe' (J. Allan Cash)

stone beds, carved-out cupboards of stone, tanks, and fireplaces. The site was discovered following a fierce gale in 1850 when the wind whipped up the sand and revealed some of the stone-built huts. Under the direction of the Laird of Skaill excavations were carried out in 1868, and the results of these aroused keen interest among archaeologists all over the world. Then in 1928–9 the late Professor Gordon Childe undertook further investigations and revealed that Skara Brae was an example of Stone Age culture. Subsequent finds suggest that the foundations of Skara Brae may have been laid as early as 2500–2000 BC.

What Eric Linklater has described as 'the master-work of all, magnificent in construction and unique in its truly megalithic grandeur', is the chambered tomb of Maeshowe,

Orkney: street scene in Kirkwall (J. Allan Cash)

3 miles from Finstown on the main road to Stromness. This has been hailed as the finest specimen of its kind in Western Europe. It is surrounded by a circular ditch and bank, 115 ft in diameter and 24 ft high. The chamber itself, which is about 15 ft square, is reached through a passage some 36 ft long and 4½ ft high. It has concealed lighting installed, and is open to the public.

Burial cells are to be found off the central chamber. All of these were empty when Maeshowe was excavated in 1861, and because of this and the fact that there were runic Viking inscriptions on the walls it was then thought that the chamber was of Norse origin. It now seems certain that it dates back centuries before this, and that the Vikings themselves removed all traces of the original occupants, as this much can be inferred from the inscriptions. One of them, for example, reads, 'The pilgrims to Jerusalem broke into Orkahowe' ('Orkahowe' being Norse for Maeshowe) – a reference to Vikings who had been on the Crusades in the 12th century. Some of the other inscriptions resemble very much modern lavatory-wall writings – 'Thorny was bedded; Helgi says so', and 'Ingigerd is the best of them all'. But other writings were much more scholarly and were obviously inscribed with care and precision, for there is one particularly long runic composition which opens with the words, 'These runes were incised by the best runester in the west, using the axe that Gauk Thrandilsson once owned in the south of Iceland.'

A smaller cairn at Onston, 2½ miles from Maeshowe, was excavated in 1884 and revealed the largest collection of Neolithic pottery ever found in Scotland – twenty-two earthenware vessels which are now housed in the National Museum of Antiquities in Edinburgh. Onston is a cairn of the 'stalled' type.

Orkney Mainland is an ever-surprising series of delights for the archaeologist. There is Winksetter in Harray parish, a farm building undoubtedly belonging to the late Norse period. The name means 'the dwelling place of Wing', and though the building has long been used as a byre and has been tampered with at various periods, it still reveals its original layout. It strongly resembles Icelandic houses of the 13th century. Records show that in the middle of the 14th century it must have belonged to Earl William Sinclair, but it certainly pre-dated this period, and is thought to have been built by one Hakon Jonsson, whose grandmother was the daughter of the Norwegian King Hakon.

A reminder of one of the great naval catastrophes of World War I is to be seen in the Kitchener Memorial Tower on Marwick Head. This was built with money raised by subscription from the people of Orkney to commemorate the death of Lord Kitchener of Khartoum, the British War Minister, who went down with HMS *Hampshire* when that cruiser struck a mine off Birsay and sank with the loss of all but about a dozen of her complement within a few miles of the Orkney shores on 5 June 1916. The cruiser was taking Lord Kitchener on a mission to Russia and her sinking inevitably gave rise to rumours of a security leak. An inquiry into the disaster was followed by the suppression of full details of the report, and subsequent investigation into this bizarre affair suggests that gross negligence on the part of the naval authorities led to the *Hampshire* sailing through an unswept channel, and that bungling by the naval staff on Orkney Mainland fatally delayed rescue operations. The Orcadians themselves were convinced that hundreds might have been saved had they not been prevented by the authorities from making rescue attempts on their own account.

The death of Kitchener was for days if not weeks regarded as a national disaster, and was certainly a blow to morale. If modern historians regard Kitchener as a military leader who was already somewhat behind the times by 1916, the fact remains that the British people of that era looked upon him as a national hero.

To the S. of Orkney Mainland lies Scapa Flow, famed as the vast anchorage which in two world wars afforded a base for the British Home Fleet. Early in the 20th century the Admiralty decided to concentrate on Orkney as a safe refuge, far from Europe, where the ships of the Royal Navy could be assembled, and it was to Scapa Flow that in July 1914 Winston Churchill, then First Lord of the Admiralty, ordered both the Home and Atlantic Fleets. Minefields were laid around the approaches and booms set up. From Scapa Flow in 1916 ships of the Royal Navy sailed to face the German High Seas Fleet in the last great naval battle of modern times involving large numbers of ships – the Battle of Jutland.

When the war was won in 1918 it was to Scapa Flow that the German Navy was ordered to sail and surrender on 23 November. There the ships remained until June 1919, awaiting the terms of the peace settlement. But the German officers were determined not to submit to any peace terms that might be devised concerning their ships. On 21 June 1919, while the British fleet was at sea on exercises, they scuttled no fewer than seventy-four of their ships. Gradually these ships were raised and salvaged for scrap metal, but it was not until just before the beginning of World War II that the last of them was finally dealt with.

If the sinking of the *Hampshire* was the greatest sensation of World War I as far as Orkney was concerned, the torpedoing of HMS *Royal Oak* inside Scapa Flow by a German submarine was undoubtedly the heaviest blow struck at the Royal Navy there in World War II. The German submarine had penetrated what was then supposed to be the safest anchorage in the world and the most closely guarded against enemy action. The *Royal Oak* sank before rescue parties could be organised and about 800 officers and men were lost.

Even Winston Churchill, once again First Lord of the Admiralty at that time (14 October 1939), paid this tribute to the enemy: 'this entry by a U-boat must be considered a remarkable exploit of professional skill and daring.'

Accommodation: there are hotels in Kirkwall and Stromness, and also some suitable for those indulging in fishing in the interior of the island. Bed and breakfast, and in some cases full board, are also provided by private houses, and many furnished cottages are let to holiday-makers in the summer. Full details of accommodation available and registered can be obtained from the Orkney Tourist Organisation in Kirkwall.

Post Offices and Banks: facilities are available.

Books: *Ancient Dwellings at Skara Brae, Orkney*, by V. Gordon Childe, 1950, and *The Bishop's Palace and the Earl's Palace*, by W. D. Simpson, 1965, both in the Ministry of Public Buildings and Works Guides and Leaflets Series, published by HMSO, Edinburgh; *St Magnus, Earl of Orkney*, by J. Mooney, Orcadian, 1935.

ORONSAY

SITUATION: in the Inner Hebrides, 1 mile S. of the island of Colonsay, to which it is connected at low tide, and 10 miles W. of the Isle of Jura.

AREA: 3 miles by 1 mile.

POPULATION: 9.

ACCESS: on foot from Colonsay at low water, or by boat from Scalasaig Pier.

It is sometimes said that this island takes its name from St Oran, a follower of St Columba, but in fact the name is Norse in origin and means 'tidal isle', which is, of course, a literal description of it.

For three hours of each day one can walk across a mile of sand to Oronsay from Colonsay, and this is just about long enough to explore the place.

The chief object of interest is the ruin of a 14th-century priory, sometimes called St Oran's Chapel. There is no doubt that Norsemen lived on Oronsay long ago, and so did some of the monks of Columba long before the priory was built. It came under the Augustinian Prior of Holyrood. Today the cloisters have fallen, but the ruins are truly beautiful, with fine carved grave-slabs which depict galleys, warriors, priests, hunting scenes, flowers, and swords. At the W. end of the re-roofed church is a magnificent high cross, tendrilled and petalled in Celtic fashion, while to the E. there is the figure of Christ holding the Host and Chalice on a small cross.

It is said that all yachtsmen and other seafarers should visit Oronsay and 'march sunways three times about the church', and then reach out an arm along a certain stone there three times 'in the name of Father, Son, and Holy Ghost'. According to ancient tradition, all who do this will never err when steering a vessel. This 'sunwise' fashion was called 'making the chessil', and the natives followed the custom in making a circuit of the island with lighted candles.

A small herd of wild goats exists on Oronsay, and like many others in the Hebrides they are said to have descended from Spanish goats which swam ashore after the wreck of the Armada.

OXCARS: *see under* Inchmickery.

OXNA

SITUATION: W. of Scalloway, Shetland Mainland.
AREA: 181 acres.
POPULATION: uninhabited.
ACCESS: by boat from Scalloway.

PABAY

SITUATION: 2 miles off Broadford, Isle of Skye.
AREA: 360 acres.
POPULATION: 12.
ACCESS: by boat from Broadford.

'A refuge for desperate men' was how Sir Donald Monro, High Dean of the Isles, described Pabay in 1549. In those days it was 'heavily wooded' and no doubt Sir Donald thought the woods made an admirable hide-out. Today it is completely treeless.

Pabay is only 95 ft above sea level at its highest point. It contains a farm-house, two holiday cottages, and various outbuildings, and has been farmed for some years. In 1967 there were various plans to set up a co-operative farming community on the island, and the Society of Friends was interested in buying it to create a 'settlement for the regeneration of spiritual and physical batteries', based on a market-gardening project.

PABBAY (Mingulay)

SITUATION: 2 miles SW. of Sandray and 1½ miles NE. of Mingulay in the Outer Hebrides.
AREA: 560 acres.
POPULATION: 6.
ACCESS: by boat from Barra or Mingulay.

Pabbay's population has declined steadily over the centuries, and in 1951 an advertisement in *The Times* offered it for sale, along with the islands of Mingulay (q.v.) and Berneray (q.v.). A prospectus then described Pabbay as having 'plenty of bracken and wild white clover . . . highly suited to cattle and would take 80 head, plus 150 sheep, without strain on grazing.'

See also Berneray.

PABBAY (Sound of Harris Islands)

SITUATION: 5 miles W. of Harris and 5 miles N. of North Uist in the Outer Hebrides.
AREA: 2 miles by 1½ miles.
POPULATION: uninhabited.
ACCESS: by boat from Lochmaddy in North Uist.

The island rises to a height of 644 ft.
See also Sound of Harris Islands.

PAPA

SITUATION: $2\frac{1}{2}$ miles sw. of Scalloway, Shetland Mainland, close to West Burra.
AREA: 148 acres.
POPULATION: uninhabited.
ACCESS: by boat from Scalloway.

Another island closely associated with the early Christian missionaries, but few traces of them remain today. In 1851 it had a population of twenty-one, declining to thirteen by 1911, after which it was evacuated.

This was one corner of the Shetlands (q.v.) where the missionary priests concentrated – at Papa Stour (q.v.), Papa Little (q.v.), and St Ninian's Isle, to the s. St Ninian's is not included in this book as, technically, it is not an island, being connected to Mainland by a narrow beach. St Ninian came to this isle a hundred years before Columba went to Iona, and it is probable that he visited Papa as well. Papa could well reveal something to archaeologists, though there is slender evidence to go on.

PAPA LITTLE

SITUATION: close to Muckle Roe and Vementry in the Shetlands, at the head of Aith Voe.
AREA: 565 acres.
POPULATION: uninhabited.
ACCESS: by boat from Shetland Mainland.

As its name suggests, Papa Little was once inhabited by the *papae*, the early Christian priests. There is little if any trace of their occupation now, but it is a pleasant and fertile island, though uninhabited. There were people here until 1850 or thereabouts.

See also Shetland Isles, Papa.

PAPA STOUR

SITUATION: 2 miles NW. of Shetland Mainland.
AREA: 3 miles by 2 miles.
POPULATION: 40.
ACCESS: by boat from Melby, Shetland Mainland.

'There are caves and caves, but probably none in the British Isles which excel those this little isle can show in weird, fantastic outline and rich colouring combined.'

So wrote John Tudor, that enthusiastic Victorian traveller and writer, of Papa Stour. And this island is indeed a spelaeologist's delight. Francie's Hole, though not very large, is a veritable fairyland: to quote John Tudor again,

> so exquisite is the colouring of the roof and sides, so pellucid is the water. What the length, breadth or height may be the writer cannot say, so overpowered with the beauty of the place was he, that he utterly forgot to estimate them. The rock forming the sides and the roof, apparently porphyritic, is partly green from sea-weed or slime, and partly red of many shades, and in places glistens like mica. The roof is studded with bosses of a deep rich purple, like the bloom of a grape, and resembling in form and regularity what are to be seen on the roofs of cathedral crypts and cloisters. Several caves branch off on the left, and at the head is a beautiful pink beach, at the top of which are alcoves or recesses like stalls in a church.

The name Papa Stour means 'the Big Isle of the Priests'. It is separated from Shetland Mainland (q.v.) by the narrow sound of Papa. Apart from the sea caves, which are probably the finest of their kind in the British Isles, it is a fertile island, especially at its E. end where most of the population is concentrated. At one time there were as many as 351 inhabitants (1871).

There is a loch here known as Dutch Loch, so called because Dutch fishermen used to land here for recreation. Nearby are some old Shetland water mills. It is an island of perpetual surprises, desolate and unattractive in bad weather and winter, but miraculously changed into an appealing, variegated isle when the sun shines and the wind drops.

See also Shetland Isles, Papa.

Accommodation: a few islanders take in visitors, but this is by special arrangement.

Book: *The Orkneys and Shetlands*, by J. R. Tudor.

PAPA STRONSAY

SITUATION: $\frac{1}{3}$ mile E. of the N. tip of Stronsay in the Orkney Islands.
AREA: 150 acres.
POPULATION: 30.
ACCESS: by boat from Whitehall on Stronsay.

It is said that the Norse warrior Earl Rognvald, having arrived in Orkney from Norway in the 11th century, and believing that he had routed his great rival, Earl Thorfinn, went to Papa Stronsay to obtain from the Culdee monks ale with which to celebrate his victory. But while carousing around a fire, Rognvald and his men were surprised by Thorfinn, who had led a counter-attack under cover of darkness. It ended in the total defeat of Rognvald's forces and his own death at the hands of Thorfinn's men.

This is one of the first known references to this small

island on which early Christian *papae*, or priests, had settled several centuries before. In their time Papa Stronsay was famed for the brewing of ale.

Once there were as many as five fish-curing stations on Papa Stronsay, and an important trade was carried on with Baltic markets. The industry revived slightly after World War I, but died after World War II.

See also Orkney Islands.

PAPA WESTRAY

SITUATION: 2 miles N. of Westray in the Orkney Islands.
AREA: 4 miles by ¼ mile.
POPULATION: 247.
ACCESS: by boat from Pierowall in Westray.

The place-name 'Papa' is frequently to be found in the Orkneys (q.v.). It derives from the Christian-Celtic *papae*, indicating the presence of priests. The Norsemen seem to have adopted the word and to have named places where Celtic clergy lived in this manner. Papa Westray is of some interest to archaeologists because there are on the island the remains of several hermits' cells. There is also a small loch in the SE. near which stand the remains of a small chapel to the memory of St Tredwall, who was blessed for dispensing fair winds to sailors. Near the chapel there is an exceptionally large weem, and 2 miles N. of the pier on the E. coast stands a natural arch.

Off the E. coast is the insignificant islet of the Holm of Papa, reputed to have been the last refuge of the now extinct great auk.

PIGMIES' ISLE, THE: *see* Luchruban.

PLADDA

SITUATION: $\frac{3}{4}$ mile S. of Arran.
AREA: $\frac{1}{2}$ mile from N. to S. and less than $\frac{1}{4}$ mile wide.
POPULATION: 6.
ACCESS: by boat from Kildonan, Arran ($1\frac{1}{2}$ miles).

Pladda is the haunt of the common and Arctic terns, and bears traces of what was once supposed to be the Chapel of St Blaise, though it could equally have been that of St Blane, who founded a monastery on Bute in the 6th century. A low-lying island, its bird life can be seen from the specially-organised cruises round the island in the summer months. A lighthouse 95 ft high is visible for 17 miles in $\frac{1}{2}$-min. group-flashing light.

See also Arran.

POMONA: *see* Orkney Mainland.

PRIEST ISLAND

SITUATION: one of the Summer Isles, $5\frac{1}{2}$ miles NW. of the entrance to Little Loch Broom, Ross and Cromarty.
AREA: 500 acres.
POPULATION: uninhabited.
ACCESS: by boat from Ullapool.

Formerly inhabited, and believed to have been occupied by one of the early Christian priests, Priest Island provides excellent fishing.

See also Summer Isles.

Pladda Island seen from Arran, with Ailsa Craig in the background (Scottish Tourist Board)

RAASAY

SITUATION: 2 miles E. of Skye and 7 miles W. of the Inverness-shire mainland.

AREA: 15 miles from NE. to SW. and about 1 mile wide.

POPULATION: 170 (1970).

ACCESS: by boat from Portree in Skye or from the Kyle of Lochalsh.

When Dr Johnson and his friend and biographer, Boswell, visited Raasay on their Hebridean tour in 1773, they were so hospitably received here, and the latter was so exhilarated, that he is recorded as having danced a jig on the summit of Dun Caan. He and the doctor were probably exploring the curious caves on the S. side of the hill. It is said locally that these caves sometimes emit smoke as a result of volcanic activity.

Dun Caan is 1,456 ft high, and is the highest point on an island of hills and dales. At the NE. of the island is the ruined castle of Brochel, home of the Macleods of Raasay, and to the SW. is Raasay House, where Johnson and Boswell were accommodated. It is now a hotel. The ever-susceptible Boswell must have been impressed by the fact that when he stayed with the Macleod of Raasay there were ten daughters in the house. Boswell declared that they 'danced every night of the year', and no doubt he danced with some if not all of them, noting somewhat wistfully that he was 'disturbed by thinking how poor a chance they had to get husbands.'

At one time Raasay had more than 600 inhabitants, but about the middle of the last century numbers began to dwindle. Today the people are mostly Free Presbyterians. One of the Macleods, the grandson of the laird who entertained Johnson and Boswell, lost almost all his money in an attempt to rescue the island from the depression into which it had sunk in the 19th century. Of all Scottish landlords of this period he was, perhaps, the most shining example of unselfish endeavour. But he failed in the attempt, and left Raasay for ever in 1846. As the poet Neil Munro has expressed it:

> Gone in the mist the brave Macleods of Raasay
> Far furth from fortune, sundered from their lands,
> And now the last grey stone of Castle Raasay
> Lies levelled with the sands.

Alasdair Alpin MacGregor tells the story of a clock tower in Raasay, the melodious chimes of which were one of the pre-1914 delights of the island. In August 1914 some thirty-six men, bound for France, fell in under the clock tower. As they marched off the clock stopped. Clockmakers from Inverness and London were called in over the years, but they failed to re-start it, and the clock has remained silent to this day.

Accommodation: 1 hotel; some cottagers take visitors.

Book: *An Island Here and There*, by Alasdair Alpin MacGregor, Kingsmead, 1972.

RABBIT ISLANDS

SITUATION: at the mouth of the Kyle of Tongue in Tongue Bay, Sutherland.
TOTAL AREA: 30 acres.
POPULATION: uninhabited.
ACCESS: by boat from Skullomie ($1\frac{1}{2}$ miles).

The group consists of three tiny islets, now uninhabited except for the creature which gave them their name – the rabbit.

RHUM

SITUATION: in the Small Isles group of the Inner Hebrides, 7 miles S. of Skye and 15 miles W. of Mallaig, Inverness-shire.
AREA: 42 sq. miles.
POPULATION: 149.
ACCESS: apply to Nature Conservancy for a permit. The Mallaig ferry boat sometimes calls at Rhum.

Rhum is roughly diamond-shaped, being 8 miles from N. to S. and about the same distance from E. to W. The Torridonian sandstone which underlies the entire island is exposed in the N., blending into smooth hills and rolling moorland. To the S. is much higher land, rising to well

Rhum and Eigg, seen from the mainland (J. Allan Cash)

over 2,000 ft and culminating in Askival, which at 2,659 ft is the highest peak on the island, closely followed by Ainsval (2,552 ft).

The Kinloch River runs E. through a beautiful glen in the centre of the island to Loch Seresort, at the head of which is the only settlement and landing place on Rhum. In 1772 there were 325 inhabitants, but in 1826 the Macleans of Coll, the owners of the island, induced all but one family to leave and replaced the traditional crofting of Rhum with sheep farming. The Marquis of Salisbury, who bought the island in 1840, converted it into a deer forest.

Forty years later John Bullough took over Rhum, and his family further restricted access and built the castle of Kinloch and the family mausoleum in Glen Harris. In 1957 the island was presented to the Nature Conservancy, and it remains a national nature reserve noted for its facilities for geological and botanical research. Permission has to be obtained from the Nature Conservancy to visit all areas except for those immediately around Kinloch.

Kinloch Castle is still furnished as it was in the Edwardian era, and is surrounded by the only trees on the island. The settlement at Kinloch possesses a post office and a small school which is also used for church services when the neighbouring minister from Eigg visits the island. To the S. of the loch one of several derelict villages can be seen. The ruins are well preserved and provide striking examples of the 'black houses' of the crofters. Measuring 35 ft by 15 ft, with rounded corners, they contain only two rooms each, with possibly an outhouse for stock and equipment. There is usually one window opposite the door, and no chimney. The fire burned in the central hearth, the smoke escaping through a hole in the roof. Few of these somewhat horrific homes are lived in today, though some have been converted and modernised.

Other examples of derelict villages may be seen at Kilmory and in Glen Harris. The mausoleum built by George Bullough is a pillared building with a timber roof, and epitomises a belated attempt to recapture something of the sombre Gothic-Romantic school. It is decorated with Italian mosaic.

An Edwardian shooting lodge in Glen Harris is now occupied by a shepherd and his family. In the NE. of the island is a disused quarry which formerly produced cornelians or bloodstones, a dark green, red-speckled variety of quartz. At one time these were much in demand for brooches and signet rings.

Some of the finest cliff scenery of Rhum is to be found at the mouth of Glen Guirdil, with views across to Canna (q.v.) and Sanday. The W. coast is navigationally treacherous and should as a rule be avoided by yachtsmen. Legend has it that the wild, golden-brown ponies of Rhum are descended from the survivors of a Spanish galleon which was part of the Armada wrecked off these coasts in 1588. The ponies are one of the great animal life attractions on the island, as are the 1,500 red deer which inhabit it. There are also considerable flocks of goats, and in the mountains golden eagles. The deer are particularly numerous in the NW. of Rhum in the area around Fionchra, the so-called 'Holy Mountain' of the isle. In the days of the Macleans of Coll this provided a sanctuary for the deer, a tradition which was maintained by the threat of death to the whole family of any person who killed deer on the mountain.

See also Small Isles.

Accommodation: there are no hotels or guest-houses on Rhum, but camping permission may be obtained from the Nature Conservancy.

Book: *The Inner Hebrides and Their Legends*, by O. F. Swire, Collins, 1964.

RISTOL

SITUATION: one of the Summer Isles, ½ mile off the Ross and Cromarty mainland, close to Alltan Dubh.
AREA: 560 acres.
POPULATION: uninhabited.
ACCESS: by boat from Achiltibuie.

Ristol was inhabited at one time, and had its own herring-curing station, linked with Ullapool.

See also Summer Isles.

ROAN

SITUATION: at the entrance to the Kyle of Tongue, Sutherland.
AREA: 1 mile by ½ mile.
POPULATION: 41.
ACCESS: by boat from Scullomie.

At high tide Roan has the appearance of being two islands. Part of it is scooped out into the form of a basin in which the soil is very fertile, and is cultivated by a few small tenants from the mainland, while its rocks are high and precipitous. The N. side abounds with deep, narrow fissures through which the wind penetrates, impregnated with saline matter with which the islanders season their fish without using salt. On the N. side also there is an arch of rock 150 ft high and 70 ft long, and in the interior is a large circular hole which is believed to communicate with the sea by a subterranean cavern.

The rocks of Roan are a fine specimen of the conglomerate which rests upon red sandstone.

ROAREIM: *see under* Flannan Isles.

ROCKALL

SITUATION: 226 miles W. of North Uist in the Outer Hebrides.
AREA: the diameter is 83 ft at normal sea level.
POPULATION: uninhabited.

Of all the islands in this book Rockall is probably the most inaccessible, but because it is part of the British Isles, though so very far away from the rest of them, it seems worthy of inclusion. It is best known for its appearances in radio weather forecasts for shipping, and as a fishing area rich in halibut.

Some 20 miles W. of St Kilda the Atlantic Ocean shelf dips sharply to 6,500 ft. It rises again at the Rockall Bank, the summit of which comprises tiny Rockall islet, reaching 70 ft at its highest point. It has a gradual slope on its E. side, while on the W. the rocks are almost vertical. Centuries ago, this granite isle was probably larger, but the ocean has long since eroded it.

One would need to sail or fly fairly close to Rockall to see it, and to land there presents tremendous problems. Absolute calm would be essential if a landing was to be made successfully. The earliest known claim of a landing was by a Captain Basil Hall in 1810, and some of his crew are reported to have secured rock specimens. Another landing was made by a British vessel in 1862, but an expedition sponsored by the Royal Irish Academy which arrived in the vicinity in 1895 was unable to land. They did, however, take soundings and made observations of the gannets, kittiwakes, puffins, and guillemots seen on the isle. It was a modern Royal Marine Commando unit which eventually landed on Rockall and succeeded in planting a Union Jack on the summit after World War II. Then in 1972 a joint Services-civilian team installed a beacon light on the island, flashing every fifteen seconds and designed to operate for a year without servicing.

Book: *Rockall*, by J. Fisher, Bles, 1956.

RONA

SITUATION: 5 miles E. of Skye and 4 miles W. of Applecross on the Ross and Cromarty mainland.
AREA: 6 miles by 1½ miles.
POPULATION: 3.
ACCESS: by boat from Skye or the Scottish mainland.

Rona's inhabitants today comprise the trio who look after the lighthouse, the beams of which can be seen for 21 miles.

Dean Monro stated that in the 16th century Rona was inhabited by 'thieves, ruggars and reivers'.

RONAY

SITUATION: 1 mile S. of the SE. tip of North Uist, 3 miles NE. of Benbecula in the Outer Hebrides, and 1 mile E. of Grimsay.
AREA: 2½ miles by 1½ miles.
POPULATION: 5.
ACCESS: by boat from North Uist or Benbecula.

This is one of a number of small isles lying in this area of the Outer Hebrides, remote and neglected compared with most of its neighbours. As the poet T. S. Muir puts it:

> O these endless little isles! and of all little isles this Ronay! Yet, much as hath been seen, not to see thee, lying clad with soft verdure, and in thine awful solitude, afar off in the lap of the wild ocean – not to see thee with the carnal eye, will be to have seen nothing!

ROUGH ISLAND (Kirkcudbrightshire)

SITUATION: $\frac{1}{3}$ mile W. of Rockcliffe in Rough Firth off the Kirkcudbrightshire coast.
AREA: 18 acres.
ACCESS: by boat, or at low tide on foot from between Rockcliffe and Colvend.

This tiny island was presented as a bird sanctuary by John and James McLellan in 1937 to the National Trust for Scotland in memory of their brother, Colonel William McLellan, of Archard Knowles.

ROUGH ISLES (Lorne), THE: *see* Garvellochs, the.

ROUSAY

SITUATION: 2 miles NE. of Orkney Mainland.
AREA: 5 miles by 3 miles.
POPULATION: 468.
ACCESS: by boat from Orkney Mainland.

A hilly island with peat-covered uplands and flagstones, Rousay contains some of the finest chambered cairns in the Orkneys (q.v.) at Midhowe and Yarso. These are believed to be communal burial-chambers of the Neolithic period. They were certainly built at a period not later than 1500 BC.

Archaeologists should certainly pay a visit to Rousay, by motor-boat from Orkney Mainland, first obtaining information from the Orkney Tourist Association in

Kirkwall if they wish to enter the cairns, which are kept locked. Midhowe has the largest of them, a dry-stone structure 105 ft long and 42 ft wide, with an outer wall 18 ft thick. It stands about 1 mile beyond Westness House, an 18th-century building sited in a small wood. Midhowe Cairn was opened in 1932 by Mr Walter Grant, guided by Dr Callender of the Edinburgh Museum of Antiquities. The central burial-chamber is about 80 ft long and 7 ft wide, and eleven pairs of large flat flagstones arranged vertically divide the place into twenty-four burial compartments. It was found that both adults and children had been buried there, and excavations produced the remains of oxen, sheep, deer, shells, and pottery.

The Ministry of Public Works has built a structure over the Midhowe Cairn to protect it from the elements. Lighting through roof windows enables visitors to inspect the interior thoroughly.

Nearby is Mansemass Hill, which provides magnificent views across the island and over the Eynhallow Sound to Eynhallow Island (q.v.). A visit should also be paid to Yarso Cairn. This yielded up to the excavators the remains of some twenty-one people, all Neolithic long-headed types, while the fragments of pottery suggested a link with a much later date.

Trumland, on the SE. corner of Rousay, has a pier which is the principal landing-place. There are other cairns in this vicinity, the most notable being the two-storeyed tomb of Taiverso Tuck, which contains two burial chambers, one above the other. Both chambers were enclosed by a circular wall. The Neolithic village of Rinyo was sited close to a farm, Bigland, S. of Faraclett Head in the NE., but practically nothing remains of this today.

Accommodation: this is limited.
Post Office.

RUM: *see* Rhum.

RYSA LITTLE

SITUATION: ½ mile E. of Hoy in the Orkneys.
AREA: 40 acres.
POPULATION: uninhabited.
ACCESS: by boat from Lyness.

A tiny island of little interest to anyone but the solitary bird-watcher, for whom there is a variety of species to observe at various times of the year.

See also Orkney Islands.

ST KILDA ISLES

SITUATION: about 34 miles WNW. of North Uist in the Outer Hebrides.
TOTAL AREA: 3¼ sq. miles.
POPULATION: uninhabited except for military personnel.
ACCESS: the National Trust for Scotland (5 Charlotte Street, Edinburgh) runs cruises around the Hebrides which take in the St Kilda Isles.

Once these four remote islands were inhabited, but after suffering many deprivations and acute poverty, the remaining islanders were evacuated in 1930. The ruins of former dwellings on the islands now provide scope for detailed study into a primitive economy.

The isles are now a bird sanctuary, leased to the Nature Conservancy by the National Trust for Scotland, who own them. They are a paradise for all lovers of sea birds, which can be found here in great numbers, especially gannets, petrels, shearwaters, fulmars, and puffins.

See also Boreray, Dun, Hirta, Soay (St Kilda).

Book: *St Kilda and other Hebridean Outliers*, by Francis Thompson, David & Charles, 1970.

ST NINIAN'S ISLE: *see under* Papa.

SAMPHREY

SITUATION: 2 miles S. of Yell and 1½ miles W. of Shetland Mainland.
AREA: 204 acres.
POPULATION: uninhabited.
ACCESS: by boat from Mossbank, Shetland Mainland.

In 1861 Samphrey had sixteen inhabitants and was partially cultivated. Ten years later not one was left. Presumably Samphrey suffered from the migration of no fewer than 3,620 people from the Shetlands (q.v.) in this decade. The island is worth exploring for some of the most unusual specimens of plant life in the Shetlands.

SANDA AND SHEEP ISLANDS

SITUATION: 2 miles S. of the Mull of Kintyre.
AREA: Sanda is 1 mile by ¼ mile; Sheep Island is about 8 acres.
POPULATION: uninhabited except for the lighthouse keepers on Sanda.
ACCESS: by boat from Kintyre.

Sanda is the larger of these two tiny islands, and has a lighthouse. Both isles are suitable for grazing.

SANDAY (Orkneys)

SITUATION: in the NE. of the Orkneys, 3 miles S. of North Ronaldsay and 1½ miles W. of Eday.
AREA: 14 miles by 1–4 miles.
POPULATION: 1,403.
ACCESS: by boat from Kirkwall (23 miles).

Sanday is a low-lying island which lacks the peat which is the traditional fuel of the Orkneys (q.v.). The inhabitants go over to the island of Eday to cut their peat and bring it back by boat.

Perhaps the most surprising thing about this island is that it provides a site for an electronics business, Sykes Robertson, which was established in 1967 entirely without government support. Today it does trade on a worldwide basis, largely due to the reputation of its founder as an innovator and original developer. It is heartening to learn that from this remote island such a firm has not only obtained contracts to build equipment in Scotland for the location of faults in the unmanned sub-stations of the North of Scotland Hydro Electric Board, but has sold language laboratories to Hong Kong.

There is a pier on the S. coast at Kettleloft, and various small settlements are scattered around the island, which is relatively fertile.

There is some slight evidence of Viking occupation in earth and stone constructions on the island.

Accommodation: there are 2 hotels on the S. coast.

SANDAY (Small Isles): *see under* Canna.

SANDRAY

SITUATION: one of the Barra Islands, ½ mile SE. of Vatersay and 2½ miles NNW. of Pabbay in the Outer Hebrides.
AREA: 1½ sq. miles.
ACCESS: by boat from Castlebay, Barra.

The least interesting of all the satellite isles of Barra, Sandray provides grazing land for sheep and little else. It is, however, the only one of the islands possessing an inland loch where there is some fishing.

SCALPA: *see* Scalpay (Lewis and Harris).

SCALPAY (Lewis and Harris)

SITUATION: 200 yds S. of Lewis and Harris at the mouth of East Loch Tarbert.
AREA: 2 miles by 1 mile.
POPULATION: 624.
ACCESS: by boat from Tarbert.

This is the largest of a group of mainly tiny and insignificant islands in East Loch Tarbert, and is sometimes spelt 'Scalpa'. It has a lighthouse, and there is some grazing land and a fair amount of bird life.

The other islands in the Loch are uninhabited, but worthy of exploration – there are almost a score of them.

SCALPAY (Skye)

SITUATION: ½ mile E. of Skye in the Inner Hebrides.
AREA: 9¼ sq. miles.
POPULATION: 6 (1970).
ACCESS: by boat from Broadford or Lochalsh.

Rising to a height of 1,298 ft, Scalpay is roughly 4 miles long and 3 miles wide. It has no roads, electricity, telephone, or shops, and the few children on the island are rowed to school on Skye each day. But it is rich in bird life and game; it has deer, including some of the red species, grouse, occasionally herons and, at the time of writing, a pair of nesting golden eagles.

The island is mainly hilly and suitable only for rough grazing and some forestry. The soil is mostly acid and peaty, with much outcropping of rock. There are several trout lochs, and approximately 200 acres of semi-arable land. The landing place is a slipway by Scalpay House at the S. end of the island. On the N. side there is one sandy bay.

Some seventy-five kinds of flora have been recorded on Scalpay, and sixty-five species of fauna, including birds. The history of the island is somewhat obscure between the period of Viking occupation and the 19th century. There is a chapel on the site of a Culdee cell or temple near Scalpay House, and this is thought to date from the 11th century. An incised standing stone still exists. Some authorities state that a Viking settlement was established here at one period, but this has not been authenticated.

In 1961 the island was purchased privately for about £25,000, including its six houses. The inhabitants include a farmer and a game-keeper. Scalpay's economy is in a state of transition. It is proposed to farm red deer on an intensive scale. About 800 acres are given over to forestry, and some sheep and cattle are kept. Trout fishing is available if permission is obtained. The owners, whose permission should be obtained before visiting the island, are Messrs A. T. and M. F. G. Walford, Invercorrie, Ullapool, Ross and Cromarty.

Accommodation (occasional): 3 cottages.

SCARBA

SITUATION: 1½ miles N. of Jura and 3½ miles W. of Argyll.
AREA: 3 miles by 2½ miles.
POPULATION: uninhabited.
ACCESS: by boat from Jura or Tobermory, but dependent on weather conditions.

Navigationally, Scarba lies in one of the most treacherous areas around the Scottish coasts. The full force of the Atlantic causes tidal races between the mainland and the islands in this vicinity. The most fearsome of these tidal races is that which forces itself through the narrow Gulf of Corrievreckan between Scarba and Jura (q.v.). The speed of the tide here can be as much as 8 to 10 knots in the westward direction.

The moor-covered isle of Scarba, with a hill rising to 1,470 ft, provides a grandstand view of the tidal races and the freak whirlpools which they throw up. What is known as the Great Whirlpool forms close under the Scarba shore at Bagh Ban. Hebrideans say of this whirlpool (which they call 'the Hag'): 'when she wears her cap one must keep away.'

Thus it is nearly always dangerous to risk a visit to Scarba, even in a fair-sized vessel. The Great Whirlpool has been known to capsize and destroy even motor yachts.

SCARP

SITUATION: 1 mile N. of Husanish Point, Lewis.
AREA: 3 miles by 2 miles.
POPULATION: 141.
ACCESS: by boat from Lewis.

Sometimes spelt Scarpay, this island rises to a peak 1,005 ft high and has a rugged grandeur, with much interesting bird and plant life. It has a tiny village on its SE. side, but a landing here is difficult when the sea is rough.

SCARPAY: *see* Scarp.

SEIL

SITUATION: 7 miles SW. of Oban, and connected with the mainland by a bridge.
AREA: 4 miles by 2 miles.
POPULATION: 369.
ACCESS: by bridge from the mainland.

Perhaps the major attraction of Seil is the curious humpbacked bridge which denies it the right of being an independent island. It is the only bridge of its kind linking any of the Hebrides to the mainland, and was designed by Telford in 1792.

Nevertheless Seil has an essentially island-minded community centred around the village of slate-roofed, white-washed cottages. It also has a slate quarry that is still working.

SEVEN HUNTERS, THE: *see* Flannan Isles.

SGEIR AM FHEOIR

SITUATION: one of the Treshnish Isles, 3 miles W. of Mull in the Inner Hebrides.
AREA: 25 acres.
POPULATION: uninhabited.
ACCESS: by boat from Achloch.

One of the smallest of the Treshnish Isles (q.v.), and a difficult one on which to land. It is said to have a colony of about forty pairs of terns, probably of the Arctic variety.

SGEIR RIGHINN: *see under* Flannan Isles.

SGEIR TOMAN: *see under* Flannan Isles.

SGEIREAN GLASG

SITUATION: one of the Summer Isles, 4 miles off the Ross and Cromarty mainland.
AREA: 7 acres.
ACCESS: by boat from Ullapool.

This island, one of the tiniest of the Summer Isles (q.v.), was bought several years ago for £50 by Mrs I. Mason, of Edinburgh, to provide her children with a hide-away holiday on a real uninhabited island. A delightful summer retreat, it must be worth far more today.

SGROT-MHOR: *see under* Monach Isles.

SHAPINSAY

SITUATION: 2 miles N. of Orkney Mainland.
AREA: 10 sq. miles.
POPULATION: 618.
ACCESS: by boat from Orkney Mainland.

Because of its proximity to Kirkwall, capital of the Orkneys (q.v.), Shapinsay is sometimes referred to as 'the suburban island'. It is the most populated of the outer Orkney Islands, and the most fertile.

The scenery is somewhat monotonous, for Shapinsay is almost all lowland. Trees never grow to a great height in Orkney, and efforts to grow them to provide a windbreak for houses have met with failure. Balfour Castle, a mansion in the Scottish Baronial style in the SW. of the island, provides an illustration of this: the trees intended to shelter it only grow a few feet high and the house towers above them.

It was Major Balfour of Balfour Castle who did more than any other man to improve the island agriculturally. He bought a large estate on the W. of Shapinsay in 1782 and transformed the whole life of the place. 'Previous to his purchase', wrote the Revd George Barry,

> nothing was to be seen over its whole extent, but a dreary waste, interspersed with arable land ill cultivated, a few miserable hovels thinly scattered over its surface . . . everything on this estate now happily wears a very different and more pleasant aspect. An elegant house has been built; the lands are substantially enclosed, and judiciously cultivated with the English plough.

The result of Major Balfour's planned improvements was a rotation of crops, and soil which before only bore coarse grass was made to produce oats, barley, pease, wheat, potatoes, clover, and turnips. A small village was built by the side of the harbour of Elwick, and scope was provided for joiners, carpenters, weavers, tailors, shoemakers, and other craftsmen. In fifty years the population increased from 642 to 730 and, said the Revd George Barry, 'the cause of this increased population we are able to trace to the residence of a single proprietor.'

Sheep-farming was also improved here by judicious cross-breeding during the régime of Major Balfour, and though the numbers have dropped, the quality has immensely improved.

Book: *History of the Orkney Islands*, by the Revd G. Barry, 1805.

SHEEP ISLAND (Kintyre): *see* Sanda and Sheep Islands.

SHEEP ISLAND (Lorne)

SITUATION: 2 miles W. of Seil Island in the Firth of Lorne, and 4 miles SE. of Mull.
AREA: ¾ mile by ¼ mile.
POPULATION: uninhabited.
ACCESS: by boat from Seil.

The island is used for grazing.

SHETLAND ISLES

SITUATION: 60 miles N. of the Orkney Islands.
TOTAL AREA: 551 sq. miles.
POPULATION: 17,843 (1966).
ACCESS: daily air service from Renfrew to Sumburgh Airport on Shetland Mainland; also by air from Kirkwall, Orkney, and by sea from Aberdeen or Leith to Lerwick.

The Shetland Isles are the most northerly part of the British Isles, and to appreciate their geographical position one must note that the nearest mainland town is not in Scotland at all – it is Bergen in Norway.

They comprise about 100 islands, many of which are little more than rocks, and only twenty of which are inhabited. The largest island by far is Shetland Mainland (q.v.), which is the most convenient centre for exploring the Shetlands.

Climatically, one can say that the Shetland summers are never very warm, while the winters are rarely very cold. Nevertheless the climate of the Shetlands is noted for its heavy rainfall, mist, and cloud. To some extent the islands are affected by the warm waters of the Gulf Stream. There are frequent violent storms throughout the year.

A hundred years ago the population of the Shetlands was double what it is today. Despite the fact that life and conditions on the islands have improved enormously over the past half-century, the decline in population continues. The younger people will not stay in Shetland. It is also probably true to say that the eviction of tenants by landlords in order to create sheep farms prior to the passing of the Crofters' Holdings Act of 1886 is something from which the islands have never recovered.

The Shetlanders are much more fishermen than crofters, and have not developed their land to the same extent as the people of Orkney. They breed sheep, cattle and, of course, the famous Shetland ponies. After fishing, their prime industry is the making of Shetland shawls and other garments from the wool of their sheep. It is said that a Shetland woman can knit a shawl that can be drawn through a ring.

It is estimated that Shetland's fishing industry is worth something more than £750,000 a year.

Shetlanders still have Norse blood in them, and they still retain many Norse names, phrases, and sayings in their everyday speech. Norse was actually spoken in Shetland up to the 18th century, and the islands were a Norse dominion until 1469. The old Shetland dialect was known as Norn.

Little is known about the coming of the Picts to Shetland: they left no written records. When the Vikings arrived in the islands it is said that they found 'two races – the Picts and the Papae [Christian priests]'. The Norsemen came to Shetland in about the 8th century, and in the latter half of the following century colonised the islands. Harald Fairhair landed in the N. of the Shetlands (Haroldswick still bears his name) and invited his ally, Rognvald of More, to become Jarl of Orkney and Lord of Shetland. The Jarls held sway over the islands for nearly 500 years, and indeed Norse law was not abolished in the islands until the early part of the 17th century.

There is much in the islands to interest the archaeologist, though perhaps not as much as in Orkney. The ornithologist is well catered for in the bird sanctuaries on Noss (q.v.) and Fair Isle (q.v.). Conditions in the Shetlands favour large colonies of sea birds, and solan geese, lapwings, eider-ducks, puffins, gulls of all kinds, kittiwakes, fulmars, and even the great skuas are plentiful. The Manx shearwater, however, is becoming rare, but on the moors the grey, ringed, and golden plovers are to be found on occasions, and of the smaller birds there are redshanks, turnstones, dunlins, sanderlings, purple sandpipers and, if you are lucky, the bartailed godwit and avocet. Perhaps the tiniest bird to be recorded here is the goldcrest, weighing less than a sixth of an ounce.

The Shetland pony is the pride of Shetland's animal life, but the islanders are equally proud of their Shetland breed of sheep. Otherwise the range of animals is extremely limited. Rabbits are a pest in some areas; hares are plentiful in other parts. There are some stoats and hedgehogs, but relatively few rats, except for the colonies of black rats in Lerwick and the bluish-black rats of Whalsay.

There are rather more flowers than one would expect in such a northerly clime. In summer the islands are carpeted with the blue of vernal squill, a flower which favours stormy coasts and heavy rainfall. Primroses grow almost everywhere, even on some cliffs. There are also some wild orchids to be found – *Orchis maculata* and the handsome purple orchid, *Orchis mascula*. Interesting also is the frog orchis, *Habenaria viridis*, which has lobed tubers and yellow-green flowers. The latter is seen on Shetland Mainland and Unst. Some of the hill plants include unusual species.

See also Balta, Bigga, Bressay, Brother Isle, Burra, Cheynies, Colsay, Egilsay (Shetlands), Fair Isle, Fetlar, Foula, Haaf Gruney, Hascosay, Hildasay, Huney, Lamba, Linga (Muckle Roe, Shetlands), Linga (Unst, Shetlands), Linga (Vaila, Shetlands), Little Roe, Mousa, Muckle Flugga, Muckle Roe, Nibon, North Havra, Noss, Oxna, Papa, Papa Little, Papa Stour, Samphrey, Shetland Mainland, the Skerries (Shetlands), South Havra, Trondra, Unst, Urie Lingey, Uyea, Uynarey, Vaila, Vementry, West Linga, Whalsay, Yell.

Books: *The Shetland Book*, edited by A. T. Cluness, Zetland Education Committee, 1967; *The Orkneys and Shetlands*, by J. R. Tudor; *Shetland Life under Earl Patrick*, by Gordon Donaldson, Oliver and Boyd; *Orkney and Shetland*, by Eric Linklater, Robert Hale, 1965.

SHETLAND MAINLAND

SITUATION: 60 miles N. of the Orkney Islands.
AREA: 378 sq. miles.
POPULATION: 13,282 (1966).
ACCESS: by air from Renfrew to Sumburgh, 27 miles S. of Lerwick; by air from Kirkwall in Orkney; or by boat from Aberdeen or Leith to Lerwick.

Mainland is the principal island of the Shetlands (q.v.) and about half of its inhabitants live in the chief town, Lerwick. The second most important township is Scalloway on the W. coast, a centre of the fishing industry.

The N. of Mainland is bleak and somewhat bare, but the W. coast is exceptionally beautiful, especially the long, narrow voes, as these arms of the sea are called. Almost in the centre of the island is its highest point, Ronas Hill (1,486 ft). There are many lochs and streams where trout (but not salmon) are abundant. Shetland Mainland offers many attractions for the fishermen, not least by reason of the fact that in many of the voes and some of the lochs fishing is free. Where charges are made they are little more than nominal, and arrangements for fishing can easily be made on arrival.

At the S. end of Mainland, not far from the airport, is Jarlshof, the remains of ancient habitations dating back to the Stone, Iron, and Bronze Ages. Of greatest interest are what are called 'wheel-houses' (2nd century AD). These are circular in shape and obtain their name from the radial walls. These radial projections stop a little more than halfway to the centre, leaving an undivided space around a hearth. In some compartments there are cupboards in the walls and stone-lined pits in the floor. Finds made as a result of excavations here include clay

opposite: *Shetland Isles: ancient fortifications at Sumburgh (British Tourist Authority)*

moulds, hammerstones, bone fabricators, beads, and pottery.

Shetland Mainland is rich in prehistoric remains. There are many mounds, cairns, and *brochs* in different parts of the island, and some stones with inscriptions in Ogham lettering. Archaeologists should visit the excellent museum in Lerwick.

The Shetland capital is a modern town owing its prosperity to its excellent harbour, one of the best in the whole area: it owes much to the shelter afforded it by the nearby island of Bressay (q.v.). Bressay Sound has known many fleets anchoring in its waters, and in 1640 was the scene of a naval encounter between the Spaniards and the Dutch. For many years the Dutch and English competed for fishing rights in these waters, until in Charles I's reign, after the English fleet had opened fire on the Dutch herring fishermen, the latter agreed to pay £30,000 for the remainder of the year's fishing, and an annual payment from then on.

Yet it was not until the 19th century that Lerwick's fishing trade was extensively developed for the benefit of the Shetlanders. Mainland's other industry, to which the shops of Lerwick bear witness with their displays, is knitting. Lerwick has a Victorian town hall with stained-glass windows, a hospital, and the only secondary school in the Shetlands. It also produces a newspaper, the *Shetland Times*.

Lerwick's most famous annual festival – Up-Helly-Aa – takes place on the last Tuesday in January. It is in effect a glorification of the Viking past of Shetland, an enacting of a pagan fire-festival, marking the end of Yule-tide and symbolising the desire for the sun to appear once more after the long nights of winter. It was a Shetland poet and song-writer, Haldane Burgess, who introduced the ritual of a Norse galley for Up-Helly-Aa. Prior to this blazing barrels of tar were rolled through the town. Now the townsmen in procession carry the Norse galley to the harbour and there fling their lighted torches into the craft until it is ablaze from bows to stern, singing

Shetland Isles: Lerwick quay (J. Allan Cash)

'The Norseman's Home' as a funeral dirge. There is almost a political flavour of revolutionary fervour in some of the verses of the song as written by Haldane Burgess:

Our galley is the People's Right, the dragon of the free,
The Right that, rising in its might, brings tyrants to their knee;
The flag that flies above us is the Love of Liberty;
The waves are rolling on.

Scalloway, Mainland's other chief town, is much older than Lerwick and its castle ruin should not be missed. It was built in 1600 for Earl Patrick Stewart, who used to hang his victims from an iron ring in one of the chimneys. Close to Scalloway is the lovely Tingwall Valley, and in Tingwall Loch is a small island approached by stepping-stones on which was formerly held the open-air parliament of the Shetlands. In the days of the Norse

Shetland Isles: Up-Helly-Aa festival in Lerwick (J. Allan Cash)

legislators this parliament was known as the Shetland Althing. Eric Linklater, an authority on Shetland, says,

> it may not be necessary to accept this tradition, for there is nothing in the law books to suggest the Norse legislators had to choose the windiest and least comfortable of places for their deliberation. More probably the islet was used for holmgang, the legal, ritualistic duel which served the purpose of trial by ordeal.

Trees are scarce in Shetland: the raging gales do not encourage their growth. There are, however, some plantations of trees near Weisdale Voe, where there is some of the best sea-trout fishing in the island.

Accommodation: there are 2 hotels in Lerwick, the Grand and the Queens, and another hotel in Scalloway. There is some other accommodation in some of the villages.

Post Offices: postal facilities on Mainland are good.
Churches: there are various Church of Scotland and Presbyterian churches, and in Lerwick a Roman Catholic Church, mainly catering for visitors.
Books: *Orkney and Shetland*, by Eric Linklater, Robert Hale, 1965; *The Shetland Book*, edited by A. T. Cluness, Zetland Education Committee, 1967.

SHIANT ISLES

SITUATION: 6 miles SE. of Lewis in the Outer Hebrides, and 14 miles N. of Skye.
TOTAL AREA: 400 acres.
POPULATION: uninhabited.
ACCESS: by boat from Lewis or Skye.

That inveterate island-lover, Sir Compton Mackenzie, once owned the Shiant Isles, a small compact group of three named islands, but for practical purposes only two, as the two larger islands are joined by a narrow stony spit.

The islands have a much lovelier name in Gaelic – Eileanan Seunta, or the 'Enchanted Isles'. They used to be inhabited, but the last family – an old man and his daughter – left at the beginning of this century. After Sir Compton Mackenzie had reconditioned a cottage on one of the islands for his own use the Shiants were bought by Mr Nigel Nicolson, who lets the grazing rights to a farmer on Scalpay who pays occasional visits to the islands.

They have a peculiar charm of their own, and through the courtesy of Mr Nicolson it is possible to obtain permission to visit them today. There is some splendid rock scenery in the islands and a wealth of bird life, including the fulmar, shag, golden eagle, oyster-catcher, herring gull, razorbill, guillemot, puffin, raven, wren, skylark, kittiwake, and many other species. A record of the fact that people still take the trouble to visit these remote islands is provided by the visitors' book kept in the cottage on the main island.

See also Eilean An Tighe, Eilean Mhuire.

Books: *Island Going*, by Robert Atkinson, Collins, 1949; *A Tangle of Islands*, by L. R. Higgins, Robert Hale, 1971; *The Highlands and Islands*, by Dr Faser Darling and J. Morton Boyd, Collins, 1964.

SHIELDAIG

SITUATION: in Loch Shieldaig, Ross and Cromarty.
AREA: 32 acres.
ACCESS: by boat from Shieldaig village.

This small, tree-covered island was acquired by the National Trust for Scotland in 1970 with money from its Coastline and Islands Fund. It is set in some of the finest loch scenery in this part of Scotland: Loch Shieldaig joins Loch Torridon before entering the sea at Fearnmore.

The original trees on Shieldaig Island are said to have been planted more than a century ago to alleviate local unemployment. The Trust's intention is to preserve the character of the island, and in co-operation with the Nature Conservancy a long-term forestry plan is being prepared.

SHILLAY: *see under* Sound of Harris Islands.

SHUNA (Appin)

SITUATION: $1\frac{3}{4}$ miles NE. of Sgeir Bhuidhe light beacon in Loch Linnhe, and 1 mile from the Argyll mainland close to Appin.
AREA: 2 miles by $\frac{3}{4}$ mile.
POPULATION: 4.
ACCESS: by boat from Appin.

At low water Shuna is much closer to the mainland, as a gravelly bank extends about $2\frac{1}{2}$ cables in its direction. Shuna is noteworthy for its views of Ben Nevis and for a ruined castle at its S. extremity.

SHUNA (Lorne)

SITUATION: 2 miles W. of the Lorne district of Argyll, lying off the entrance to Loch Melfort, and 1 mile E. of Luing.
AREA: $2\frac{1}{4}$ miles by $1\frac{1}{2}$ miles.
POPULATION: 8.
ACCESS: by boat from the Scottish mainland or from Luing Island.

The island rises to a height of 296 ft. Much of the land is suitable for grazing.

SILLAY: *see under* Monach Isles.

THE SKERRIES (Shetlands)

SITUATION: 5 miles NE. of Whalsay in the Shetland Isles.
TOTAL AREA: 1 sq. mile.
POPULATION: 108.
ACCESS: by boat from Isbister in Whalsay or from Lerwick, Shetland Mainland (weekly steamer service).

The Skerries, unlike most of the Shetlands (q.v.), have actually increased their population over the past century. In 1825 there were only seventy-five inhabitants. The reason for this is that the Skerries have an exceptionally fine fishing ground all around them, as well as possessing a good natural harbour. Fishing is the mainstay of the Skerries.

They comprise several tiny islets, on one of which, Bound Skerry, which is not much more than a large rock, there is a lighthouse. The two inhabited islands, Housay and Bruray, are linked by a bridge. One of the assets of the Skerries is that they are better off for communications and transport than any other remote island or island group in the Shetlands. Not only are they easily accessible, even when the weather is bad, but the steamer service from Lerwick is an excellent one.

SKYE

SITUATION: most northerly of the Inner Hebrides, separated from the mainland of Inverness-shire by the Sound of Sleat, and only 300 yds at its nearest point from the Kyle of Lochalsh.
AREA: 535 sq. miles.
POPULATION: 8,000.
ACCESS: by steamer or car ferry from the Kyle of Lochalsh to Kyleakin.

Skye is the romantic's island, a place of lofty mountains, lovely glens, mist and mystery, on occasions illuminated either by the most magnificent rainbows or by sunsets of technicolour radiance. 'To visit the island', wrote Alexander Smith in *A Summer in Skye*, 'is to turn your back on the present and walk into antiquity . . . and, more than all, the island is pervaded by a subtle, spiritual atmosphere.'

It is the largest of the Western Isles, 50 miles long and roughly 20 miles across on average. Its loch-indented coast means that no point on the island is more than 5 miles from the sea. In spite of its largely mountainous character much of the land is fertile and provides an adequate living for the crofters who form the bulk of the population.

Perhaps the finest scenery in the island is to be found in the Cuillins, which stretch from Loch Brittle on the W. to Broadford on the E., a line which runs just S. of the centre of the island. They are divided by Glen Sligachan into the Black and Red Cuillins, the former (in the W.) being composed of a hard rock, gabbro, otherwise found only in the N. of Norway. Here the mountains rise to above 3,000 ft. At the S. end of the range, in Glen Brittle, is one of the finest waterfalls in the British Isles. The Red Cuillins (to the E.) are so called because the granite gives off a pink glow in certain lights: they are lower and more rounded than their western neighbouring peaks.

Sligachan, situated halfway between Portree and

Skye: Portree (Radio Times Hulton Picture Library)

Broadford, is the best base for exploring the Cuillins. It has a hotel used by rock climbers and anglers, the fishing in the nearby river being excellent. This village is pleasantly situated at the head of Loch Sligachan against a back-drop of mountains.

Portree ('Port of the King') is the chief town of Skye. Named after a visit by King James V of Scotland, this fishing harbour accounts for more than 2,000 of the population of Skye. It is a pleasant centre for visitors, with its views across to the island of Raasay and its white-washed houses. To its N. lies the Trotternish Peninsula, culminating in the mountain of Quirang (1,800 ft), which has a precipice on its NE. side of columnar and fluted basalt rocks standing out from its face. Quirang has been

opposite: *Skye: Loch Scavaig (British Tourist Authority)*

described as 'Britain's most spectacular mountain', with its fantastically-shaped pillars and gullies and remarkable geological structure. Also on the W. of Skye, N. of Portree, is the Old Man of Storr, a pinnacle composed of solidified lava which cooled at the mouth of a now-extinct volcano. For centuries this pinnacle remained a challenge to climbers which was not taken up. It was not in fact climbed until 1955, when Don Willan led a party of three to the top.

Skye can also boast the most impressive sea cliffs in the Hebrides. These extend down the NW. coast. Particularly impressive are those at Dunvegan Head which reach to heights of up to 1,000 ft. Dunvegan, however, is better known for its castle, the seat of the chiefs of the MacLeod clan. Part of the castle dates back to the 10th century, with various renovations and additions carried out at later dates. No other castle has remained in the hands of one family for such a length of time.

The castle is moated, and within the 10-ft-thick walls it contains many relics of past history, including Rory Mhor's two-handed sword, his drinking horn, the 'Fairy Flag' believed to have been captured from the Saracens during a Crusade, and a lock of the hair of 'Bonnie Prince Charlie'.

Dunvegan is one of five castles on the island. The other four, though in ruins, are all worth visiting. Two of them, Maol at Kyleakin and Dunscairth at Sleat, are supposed to have been built by warrior-queens. Skye has quite a history of feminine heroism, for at a third castle, Camus, situated opposite Dunscairth on the E. side of the Sleat Peninsula, 'Mary of the Castle' MacDonald's courage saved the day in battle against the MacLeods. Later still, of course, another MacDonald, Flora, helped Prince Charles Edward Stuart 'over the sea to Skye'.

As its name suggests, Skye was inhabited by the Norsemen for a lengthy period. In 1263 disputes between the islanders and their Norse overlords precipitated the attack upon Scotland by King Hakon of Norway. Hakon's fleet was anchored at Kyleakin before being

Skye: Needle Rock, Quaraing (Radio Times Hulton Picture Lib.)

defeated by Alexander III of Scotland at the Battle of Largs. This battle resulted in the Norsemen finally abandoning their claims to the Western Isles, and after this Skye was ruled by the MacLeods and the MacDonalds of Sleat, who were continually feuding with one another.

The MacDonalds ruled the northern and southern peninsulas of Skye, but their participation in the 1715 Rebellion led to a speedy decline in their fortunes. Both the MacDonalds and the MacLeods learnt their lesson from the Old Pretender's failure in this revolt, and in consequence neither took part in the 1745 Rebellion.

Dunscairth, which was the legendary dun of Sgathach, the warrior-queen of Skye in Fingalian times, was a possession of the MacDonald chiefs for some centuries, but in the 17th century they gave this up for Duntulm Castle at the N. end of the island. Duntulm, a prominent if gaunt landmark on a forbidding headland, fell into disrepair as the MacDonalds' fortunes ebbed, and about 1732–3 it was finally abandoned. Lack of money was the most likely reason for this, but there are stories that the occupants were driven out because of the frequent visitations of an ancestral ghost, Donald Gorm Mhor MacDonald.

It was a later MacDonald, Flora, however, whose memory was to outlast that of any of the male members of the clan. While the chiefs of both clans lay low during the 1745 Rebellion it was Flora MacDonald who came to the rescue of Prince Charles Edward Stuart after the defeat at Culloden.

Flora, who was living in a cottage on South Uist (q.v.) at the time the Prince was fleeing from the Government troops, obtained passes for herself and a female servant to go to Skye. The Prince was disguised as the servant woman in a flowered linen gown, adopting the name of Betty Burke. Bishop Forbes in *A Lyon in Mourning* asserted that the Prince's attitude to Flora was one of the utmost deference: 'When she came into the room, he always rose from his seat and paid her the same respects as if she had been a queen.'

From South Uist they arrived off the NW. corner of Skye, but were fired on from the beach and rowed on to Kilbride in Trotternish, near the home of MacDonald of Kingsburgh, where the Prince was sheltered. Before he left on the next stage of his escape the Prince, 'laying his arms about her [Flora's] waist, and his head upon her lap, desired her to cut out the lock [of his hair] with her own hands.' This she did, giving one half to Mrs MacDonald of Kingsburgh, and keeping the other for herself.

Years later when Dr Samuel Johnson and his biographer, Boswell, visited Skye (1773), the doctor met Flora MacDonald and, it is recorded, they 'quite liked one another'. Johnson testified eloquently to the hospitality of the people of Skye. The ruins of Coirechatachan, where Mistress MacKinnon embraced Johnson with the words, 'What is it to live and not to love?', may still be seen near Broadford.

While on the subject of the MacKinnon family, it is worth while recording the stories of the origin of that famous liqueur associated with the 1745 Rebellion, drambuie. It has been said that drambuie was first drunk on Scottish soil on the isle of Eriskay (q.v.) when the Prince first landed there, the secret recipe for the liqueur having been brought over from France.

The history of drambuie is told in a small pamphlet issued by the Drambuie Liqueur Company, Ltd as follows:

> In 1745 when Prince Charles Edward made his gallant but unsuccessful attempt to regain the throne of his ancestors, the MacKinnons of Skye were able to play a prominent part in assisting the Prince to evade his pursuers. In token of his gratitude, one of the family was presented with the secret recipe of the Prince's personal liqueur. The secret is held by the same MacKinnon family to this day. When you open a bottle of Drambuie, you are drinking the identical liqueur that was once the favourite of the last of the Stuarts.

In the late 18th century the harsh policy of the clearances forced many crofters to emigrate, but Skye did not suffer as badly as other islands in this respect. Indeed in 1882 the resistance of large numbers of crofters to the Sheriff's Officers led to a commission of inquiry being set up by Gladstone, then Prime Minister. The outcome of this was the Crofters' Holding Act of 1886, which did a great deal to ease the lot of the Highlanders both on the mainland and in the islands. Incidentally, the women of Skye were active in preventing their menfolk's eviction. While their husbands were at work they kept off the officials by the use of stockings filled with stones.

Skye has rich agricultural land, every bit as good as is to be found on the mainland, and farming is the chief occupation of the inhabitants. It is admirable terrain for adventurous rock climbers and for fishermen (its brown trout and salmon are prolific in the various lochs), and there is a wide variety of animal and plant life. Hawks

Sunset over Skye (Radio Times Hulton Picture Library)

and tawny owls abound, and grey seals bask on the rocks of the SW. shore. Occasionally red deer may be seen. For a visitor the weather is the main hazard. July and August on average are the wettest months, while June and September are usually sunny.

Some of the lochs in the interior of the island are to be found at surprising heights like Loch An Fhirbhallaich, 950 ft above sea level. Glen Brittle is probably the best centre for both walkers and climbers. The Forestry Commission have planted trees extensively in this area.

Twenty miles from Portree is Kilmuir, the burial place of Flora MacDonald. It is marked by a 28-ft memorial cross, one of the largest of its kind in Scotland.

Skye has a reputation for producing military leaders. It is recorded that in the Napoleonic Wars the island provided the British Army with seventy-one generals and colonels, and 600 other commissioned officers, as well as 170 pipers and 10,000 foot soldiers. It should be noted that in 1821 the island's population was more than 20,000. Even one of Napoleon's marshals was a MacDonald!

Accommodation: there are hotels at Sligachan, Glen Brittle, Portree, Kyleakin, Dunvegan, and Broadford. There is also a considerable amount of accommodation available at cottages and boarding-houses. For further information apply to the Isle of Skye Tourist Organisation, Area Tourist Office, Meall House, Portree.

Post Offices and banks: at all the chief towns.

Books: *Description of the Western Isles of Scotland*, by Dr John MacCulloch, 1819; *The Enchanted Isles*, by Alasdair Alpin MacGregor, Michael Joseph, 1967; *A Summer in Skye*, by Alexander Smith, 1865; *Climbing Guide to the Cuillin of Skye*, Scottish Mountaineering Club, 1958; *The Island of Skye*, Scottish Mountaineering Club, 1954; *The Journal of a Tour to the Hebrides with Samuel Johnson*, by James Boswell, 1773; *Journey to the Western Islands of Scotland*, by Dr Samuel Johnson, 1775; *The Prince in the Heather: the Story of Bonnie Prince Charlie's Escape*, by Eric Linklater, Hodder and Stoughton, 1965.

THE SMALL ISLES

SITUATION: S. of Skye and 6 miles NNW. of Ardnamurchan in Inverness-shire.
TOTAL AREA: $59\frac{2}{3}$ sq. miles.
POPULATION: 314.
ACCESS: by David MacBrayne steamers from the Kyle of Lochalsh or Mallaig.

This group comprises the islands of Rhum, Eigg, Canna, and Muck, each entirely different from the others and all well worth a visit. There is no regular service to Muck in winter time, however.

See also Canna, Eigg, Muck, Rhum.

SOAY (St Kilda)

SITUATION: in the St Kilda group about 34 miles WNW. of North Uist in the Outer Hebrides.
AREA: 244 acres.
POPULATION: uninhabited.
ACCESS: by boat from Oban.

A strait about a quarter of a mile wide separates Soay from Hirta (q.v.), the principal of the four St Kilda Isles (q.v.). It is circumferenced by steep cliffs, but its surface is covered with green turf on which graze the unique breed of Soay sheep, believed to be descended from the wild mouflon. They live in a state of wildness.

Soay rises to a height of 1,225 ft and its only safe landing place is at the SE. corner of the island. There is a curious structure on the island, reputed to have been covered originally with turf, and it is believed to have been once upon a time an altar.

The name Soay is derived from the Norse *saudhr*, meaning 'island of sheep'. In 1764 the Revd Kenneth MacAulay, sent to these islands as a missionary, reported that there were 500 sheep on Soay and said they were difficult to catch. They were said then to be hunted by the St Kildans.

The Soay sheep are short-haired and long-legged, and have a goat-like movement, their horns lifting high from the head and having a wide curve.

opposite: *Soay (Aerofilms)*

SOAY (Skye)

SITUATION: 1 mile S. of Skye at the entrance to Loch Scavaig and 11 miles NE. of the Sound of Canna.
AREA: $3\frac{1}{2}$ miles by 1–2 miles.
POPULATION: 51.
ACCESS: by boat from Skye.

Soay was the name given to small islands where sheep were grazing by the Norsemen when they arrived off the Scottish coast, and usually refers to a small island lying off a larger one. This Soay is a mainly flat island with a number of small ponds in its interior, though it has one hill, called Beinn Bbreac, which rises to 435 ft. A narrow inlet is (somewhat pretentiously) called Soay Harbour, and this was once the centre of a shark-fishing industry. The factory connected with the industry still remains beside the derelict pier.

Gavin Maxwell, the author, was once an inhabitant of Soay – he lived here with his family in the old manse on the narrow isthmus which joins the NW. and SE. ends of the island. There are a number of derelict houses on this once-extensively-occupied island, and a few cottages now owned by artists or men engaged in lobster fishing.

It is possible to walk round the island in a single day's visit and to enjoy the display of fuchsias and other flowers that bloom there.

SORAY: *see under* Flannan Isles.

SOUND OF HARRIS ISLANDS

SITUATION: extending from NW. to SE. across the 10-mile Sound of Harris and between Harris and North Uist in the Outer Hebrides.
TOTAL AREA: $3\frac{1}{2}$ sq. miles.
POPULATION: mainly uninhabited.
ACCESS: by boat from Harris or North Uist.

Many of these tiny isles are only a few acres in size, and some little more than rocks. The principal ones are Pabbay, Boreray (not to be confused with the St Kilda Boreray), Harmetray, Groay, Kelligray, Ensay, Lingay, and Gilsay. Shillay in the extreme NE. of the group has some interesting bird life, while the islands around Harmetray in the SW. are favoured for fishing.

See also Pabbay (Sound of Harris Islands).

SOUTH HAVRA

SITUATION: 2 miles S. of Burra island and 2 miles W. of Shetland Mainland.
AREA: 147 acres.
POPULATION: uninhabited.
ACCESS: by boat from Shetland Mainland.

Up to World War I South Havra was inhabited, its population over a century fluctuating between forty-five and twenty-five. It is a pleasant island, with a great many sea birds, fair grazing land, and in parts land suitable for cultivation.

See also Shetland Isles.

SOUTH RONALDSAY

SITUATION: the most southerly of the Orkneys, ¼ mile S. of Orkney Mainland.
AREA: 7 miles by 2 miles.
POPULATION: 1,545.
ACCESS: by the short sea ferry route across the Pentland Firth, or by the road bridge link with Orkney Mainland.

South Ronaldsay is, in the strict technical sense, no longer a separate island; like Burray (q.v.) it is joined to Orkney Mainland (q.v.) by the road route across the Churchill Barriers. These were erected as part of the defences of the approach to Scapa Flow during World War II, and were later used to provide a road link which undoubtedly saved South Ronaldsay from further depopulation. But the island is included in this book partly because it is so easily accessible, and partly because it has some of the finest cliff views in the Orkneys. A walk around the island along the cliff tops is well worth the effort, not least for the ornithologist.

At the N. end of the island is St Margaret's Hope, where the ship on which Queen Margaret, the Maid of Norway, died while on her way to Scotland put in and received Bishop Dolgfimur on board before returning to Norway (1290). St Margaret's Hope is the principal settlement and is slowly regaining some of the sea-borne trade it possessed centuries ago. The island suffered considerably from the curse of bad landlords typical of the Orkneys in the 19th century, when tenants were often forced to operate small plots that could not possibly provide a livelihood in order to drive them to boost kelp production for the landlords' benefit.

See also Orkney Islands.

SOUTH UIST

SITUATION: in the Outer Hebrides, linked by a causeway to Benbecula, due N., and 4½ miles NNE. of the Barra Isles.

AREA: 141 sq. miles.

POPULATION: 2,500 (1966).

ACCESS: via Benbecula, to which there are air services from Glasgow and elsewhere; or by boat from Mallaig, Inverness mainland, and Barra to Lochboisdale. There is also a boat direct from Oban to Lochboisdale on Mondays only.

South Uist offers a wide range of scenery and attractions, from the mountains and sea lochs of its E. coast and the interior of peat bogs and excellent fishing lochs, to the beaches on its W. coast. It is also the home of the Army's rocket range for guided missiles, which was established there in 1955 after considerable controversy and correspondence in *The Times*. The activity of the rocket range has declined somewhat in recent years, and has made little difference to the life of the island which, despite the range, possesses a nature reserve.

In fact the island seems to have the best of all worlds, something which can rarely be said for any island in the Outer Hebrides. The rocket range has brought some economic advantages, even if they are marginal; it is a fairly prosperous crofting territory; and a certain benefit is derived from tourism, limited though that may be. It has the added advantage of being the largest of the three islands (Benbecula, North Uist, and South Uist) and is also linked to the other two by air, sea, and causeways. It has the two highest mountains of the Uists in Beinn Mhor (2,034 ft) and Hecla (1,988 ft), situated in the N. part of the island.

In this northern area there are no habitations, but there are the traces of ancient settlements, notably the earth houses on the lower part of Hecla. The cliffs in this part of South Uist rise to 850 ft and present some challenging posers to rock climbers.

It was in Corodale Bay on the E. coast that Prince Charles Edward Stuart landed after his flight from Culloden in 1746. The ruin of Flora MacDonald's cottage at Milton is one of the historical attractions of the island, and the story of her rescue of the Prince belongs as much to this island as it does to Skye. It was in this part of the world that he wandered from one hiding-place to another before eventually he escaped to France.

There are facilities for fishing (the island has 190 lochs, mostly filled with brown trout), sea angling, and golf on the island. Lochboisdale is the main village and port, and permission to angle can be obtained here from the factor of the South Uist estates. This is a fisherman's paradise and the fishing costs only about 40–50p a day, including the use of a boat. Salmon and sea-trout fishing costs about £3·20 a day plus £3 a day for a gillie.

The nature reserve is at Loch Druidebeg and includes several heronries and a breeding ground for the greylag goose. There is an island folk museum which is filled with examples of local arts and crafts.

Most of the inhabitants are Roman Catholics, and there are four Catholic churches and one Church of Scotland. One outstanding monument that must be seen is the 30-ft-high statue of 'Our Lady of the Isles', sculptured by Hew Lorimer in 1954, and situated on a high rock at the N. end of the island.

Today's improved standards of living on South Uist are in stark contrast to those described by Miss Ada Goodrich Freer in her book on the Outer Hebrides, published in 1902: 'Nowhere in our proud Empire', she wrote,

> was there a spot more desolate, grim, hopelessly poverty-stricken, a wilderness of rock and standing water, on which, in summer, golden lichen and spreading water-lilies mocked the ghastly secrets of starvation and disease. The huts of the people, inconceivably wretched, hard to distinguish from the peat-stacks beside them, were built of undressed stones piled together without mortar and thatched with turf.

Accommodation: there is a hotel at Lochboisdale and 2 inns, one at each end of the island. Many crofters take in visitors.

Books: *Lonely Isles, being an account of Several Voyages to the Hebrides and Shetland*, by R. Svensson, Batsford, 1955; *The Scottish Islands*, by G. Scott-Moncrieff, Oliver & Boyd, 1965.

SOYEA

SITUATION: at the mouth of Loch Inver, 1¾ miles off the Sutherland coast and 3 miles W. of Lochinver.
AREA: ¾ mile by ¼ mile.
POPULATION: uninhabited.
ACCESS: by boat from the Scottish mainland.

Soyea is mainly rock and heathland, comprised of Lewisian gneiss and Torridonian sandstone. Its highest point is 110 ft. The owners are the Assynt Estate Office, Lochinver, Sutherland.

SROGAY: *see under* Monach Isles.

STAC AN ARMIN: *see under* Boreray (St Kilda Isles).

STAC LEE: *see under* Boreray (St Kilda Isles).

STAFFA

SITUATION: 8 miles W. of Mull in the Inner Hebrides and midway between that island and the Treshnish Isles.
AREA: 70 acres.
POPULATION: uninhabited.
ACCESS: by boat from Mull; excursion steamers from Oban pass close to the island in summer.

This basaltic outcrop in the Inner Hebrides has attracted many famous visitors who have been fascinated by its majestic beauties, including Wordsworth, Keats, Tennyson, and Mendelssohn. The last-named was so attracted by the natural 'architecture' of the celebrated Fingal's Cave on the island that he called it 'the cathedral of the sea', and it inspired him to write his 'Hebrides' overture.

The island, which rises to 135 ft at its highest point, was shaped by volcanic action, and its cliffs and caves do constitute one of nature's most magnificent and aesthetically pleasing works of art. Groups of basalt columns in the entrances to some caves give the appearance of the pipes of a gigantic organ. But it is the noise of the Atlantic rollers crashing against the rocks which provides the symphony of sound, as Mendelssohn was quick to appreciate.

Staffa first became famous after a group of scientists *en route* for Iceland called in at the island in 1772. Sir Joseph Banks, who was a member of the party, made a geological report on Staffa, and subsequently this and other details were published in a book by that tireless traveller Thomas Pennant. From then on it became a fashionable pilgrimage for poets, writers, and musicians to visit Staffa. What Mendelssohn did for the island in the sphere of music Turner accomplished on the easel.

The island has not been inhabited for more than a century, and it has only been in the hands of five families in eleven hundred years. In the past it is said to have produced substantial crops of corn, but there are certainly

Staffa: sea off Fingal's Cave (Radio Times Hulton Picture Lib.)

no indications today that crops could be grown here, though it provides grazing for cattle in summer. One of the families who have held Staffa was that of the Formans, who originally won it by a bet in 1821. Captain Gerald Newell bought Staffa for £10,000 in 1968, but put it on the market again in 1971.

In an interview in 1971 Captain Newell stated that the island had its own stamp, which provided an income. 'We have 15p and 5p stamps which are very popular with tourists and collectors,' he said. 'They buy the stamps on the island, post them at a mail box there, and we take them across to the mainland. Post Office stamps are put on letters there for which the tourists have already paid. The stamps show Mendelssohn and Queen Victoria. At the moment we are getting most of our inquiries from Americans.'

Any visit to the island will be as rewarding for geologists as for those more artistically inclined. The columnar structure is relatively rare in the British Isles and Eire: it is to be found only in nearby Ulva, in the region around Ard'eanach, and in Bloody Bay near Tobermory. The cooling off of the lava sheets after volcanic action occurred irregularly and at different rates, leading to polygonal jointing, and hexagonal and even octagonal columns were formed. This has resulted in the magnificently-curved columns on Staffa at the Clamshell Cave and the Buchaille Rock and the organ-pipe pillars in front of Fingal's Cave. This cave is 66 ft high, 227 ft deep, and a mere 42 ft in width at the entrance. It is possible to follow the cave path for a considerable distance along the tops of the wave-cut basalt columns.

The columns of rock which typify Staffa gave the island its name, which in Norse means 'Pillar Island'. Equally interesting, and showing how the ancients anticipated Mendelssohn, is the origin of the name Fingal's Cave. In Gaelic it means 'the cave of melody', referring not merely to the sound of the sea birds but to the echoing of the crashing waves inside the cave. It must be remembered that when Mendelssohn was at the height of his powers the romantic period in the arts was at its peak. He was one of the first to compose independent concert overtures, and his 'Fingal's Cave', or 'Hebrides' overture, is a perfect example of the romantic temper in music.

There are other caves of interest on the island, notably Clamshell, which has basalt columns shaped in grotesque curves; the Cormorants' (or McKinnon's) Cave, 224 ft deep; and the Boat Cave, 150 ft deep. McKinnon's Cave is named after an Abbot of Iona, who is said to have moved his cell away from Staffa because of the distracting roar of the waves.

Queen Victoria, who landed on Staffa with Prince Albert in 1847, was particularly entranced by the island, and recorded in her diaries her 'delight' in all she saw. Visitors should be warned that in 1968 the landing on

Staffa of the daily cruise ship from Oban was discontinued mainly because of the reported danger of falls of rock in Fingal's Cave. The cruise ship still goes close in to the island, but those wanting to step ashore on Staffa will have to make special arrangements on Mull – a motor-boat goes to Staffa from Ulva Ferry.

There are various sea birds on Staffa – razorbills, guillemots, wild duck, puffins, and oyster-catchers.

Books: *The Isle of Mull*, by P. A. MacNab, David & Charles, 1970; *A Tour in Scotland and Voyages to the Hebrides*, by Thomas Pennant, 1772; *Leaves from a Journal of Our Life in the Highlands*, by Queen Victoria, Smith Elder & Co, 1868.

STROMA

SITUATION: in the Pentland Firth, half-way between the Orkneys and the coast of Caithness, Canisbay being about 3½ miles to the S.
AREA: 3¾ miles by 1½ miles.
POPULATION: 6.
ACCESS: by boat from Thurso.

Rows of empty, yet still habitable houses make Stroma a ghost island. Fifty years ago there were 300 people living there and 100 children at the school. Ever since the population has been falling. In 1959, when 130 people were left on Stroma, a new harbour was built at a cost of £30,000 in a brave effort to prevent the population from falling further. Today it is empty and useless, for the remaining inhabitants actually speeded up their departure after the harbour was built.

Some years ago a controversy raged when it was announced that Stroma was to be offered as a prize in an American television programme contest. The idea was abandoned after protests and barbed comments by Lord

Lyon King of Arms. Mr James Simpson, junior, of Mey, Caithness, who had left the island seventeen years earlier, bought Stroma after the fuss had subsided.

STRONSAY

SITUATION: 4 miles E. of Shapinsay and 13 miles NE. of Kirkwall, Mainland, in the Orkney Islands.
AREA: 7 miles by ½–6 miles.
POPULATION: 825 (1964).
ACCESS: by boat from Orkney Mainland.

The area around Stronsay is particularly good for fishing for all white fish, lobster, and crab. Crofting was developed here in the 18th century on what had until then been waste land, and by the middle of the following century it was reported that 100 acres of waste land had been put to the plough. But while crofting has declined, the fishing industry had a revival in the 1960s.

Whitehall on the NE. end of the island is the fishing centre and the principal settlement.

See also Orkney Islands.

SULASGEIR

SITUATION: 12 miles WSW. of North Rona and about 25 miles N. of Lewis in the Outer Hebrides.
AREA: ½ mile long, with a maximum breadth of 200 yds.
POPULATION: uninhabited.
ACCESS: by boat from Stornoway.

Sulasgeir was inhabited in the distant past, but there is

no record of any population after the middle of the 16th century. At the S. end of the island there are the remains of a chapel or anchorite's cell and some stone shelters which suggest earlier habitation. It is a bleak place, and probably it was for this reason that, centuries ago, it was selected as a site for a prison for sheep-stealers from Lewis. It is even suggested that Sulasgeir was where Brenhilda, the sister of St Ronan, was once imprisoned.

Perhaps for this reason, and because of the Highlanders' traditional sentimentalism about the past, the remains of the chapel and shelters have always been carefully preserved, and at periods even repaired by visiting fishermen and fowlers from North Ronay (q.v.) and Lewis. No doubt this reverence for the past was enhanced when the Young Pretender, 'Bonnie Prince Charlie', was greeted by a boat near Sulasgeir before he landed on the mainland.

The many tiny islets and rocks surrounding Sulasgeir include Thamna Sgeir to the E., Boghannan s'Iar, Bogha Leathainn, Da Bogha Lamha Cleit, and Mamha Cleit to the W., and Gralisgeir to the S. Much further out to the NW. is Bogha Corr. All the islands are breeding grounds for gannets, puffins, kittiwakes, and fulmars. The main vegetation is grass, mayweed, thrift, orache, and chickweed. The highest point of Sulasgeir, at the S. end of the island, is about 230 ft. The middle of the island is low-lying, and waves break right across it in rough weather. There is also a cave which runs from one side of the island to the other.

Robert Atkinson, writing of a night he spent on Sulasgeir in 1938, described how the ground 'churned with petrels, they scattered underfoot, they were headlong in the air . . . all day the sky full of gannets and their unceasing cacophony; a brief no-man's-land of silence at dusk and dawn; all night the wild dashing and outcry of petrels, flying out over the heads of the sleeping gannets.'

Sulasgeir was made a Nature Reserve in 1956.

Book: *Island-Going, to the Remoter Isles, Chiefly Uninhabited, of the N.W. End of Scotland*, by Robert Atkinson, Collins, 1949.

SULE SKERRY

SITUATION: 30 miles W. of Orkney Mainland.
AREA: 3 acres.
POPULATION: 3.
ACCESS: by special permission from the Scottish Lighthouse servicing vessels.

Sule Skerry is always perilous of access and dependent on the weather. Normally it is visited by a lighthouse servicing vessel every three weeks, and a relief lighthouse keeper is put ashore. It is a bleak, rock-bound island covered with guano. Stores are loaded on to a bogie and hauled up to the lighthouse by means of a cable fixed to a petrol-driven drum.

See also Orkney Islands.

THE SUMMER ISLES

SITUATION: extending from ½ mile off the coast of Ross and Cromarty close to Alltan Dubh to Isle Martin at the entrance to Loch Broom.
TOTAL AREA: 4 sq. miles.
POPULATION: virtually uninhabited, though there may be occasional and seasonal occupants.
ACCESS: boat trips to the islands operate from Ullapool in Loch Broom and Achiltibuie.

The Summer Isles are strung out in two lines, one running from Ristol in the N. to Priest Island in the SW., the other extending fairly close to the coast in a SE. direction from Ristol to Isle Martin. There are some twelve islands and, as their name suggests, they are to be seen at their verdant best in summer, when their flower and plant life is blooming.

The Summer Isles (Aerofilms)

Though virtually uninhabited, they still have a few visitors during the season, and some of the isles are still used for sheep grazing. There is also excellent sea fishing all around the Summer Isles. In the 19th century, when the islands were partially inhabited, curing stations were set up on some of them. In the early part of the century herring were plentiful around here and in Loch Broom, but catches declined before the middle of the century and the curing industry ended in these parts.

Seals are to be seen breeding on some of the islands, and the bird life is particularly varied.

See also Bottle Island, Carn Isles, Eilean Choinaid, Eilean Na Saille, Horse Island (Summer Isles), Isle Martin, Mullagrach, Priest Island, Ristol, Sgeirean Glasg, Tanera Beg, Tanera More.

Accommodation: there is said to be limited holiday accommodation available during the summer, but details must be obtained at Ullapool.

Book: *Island Years*, by Dr Fraser Darling.

SWITHA

SITUATION: $1\frac{1}{2}$ miles S. of Flotta in Scapa Flow, Orkneys, at the entrance to Hoxa Sound.
AREA: 50 acres.
POPULATION: uninhabited.
ACCESS: by boat from Flotta.

A low-lying island where the storm-petrel burrows and nests. Its most prominent landmarks are the Standing Stones, henge monuments of a type to be found in various parts of Orkney, and evidence of Bronze Age settlements.

See also Orkney Islands.

SWONA

SITUATION: 3 miles W. of South Ronaldsay in the Orkney Islands.
AREA: 2 miles by $\frac{1}{2}$ mile.
POPULATION: 3 (1968).
ACCESS: by boat from Thurso on the Scottish mainland (24 miles).

Once Swona had fifty inhabitants; the sole remaining trio, two brothers and a sister, were all born on the island and have been alone there for more than forty years.

They make a living out of cattle, sheep, and fishing, but admit that to do so is getting harder all the time, especially with rising costs.

There is not much to see on Swona except for sheep, and the crossing over the turbulent Pentland Firth is hardly worth while except on the calmest of days.

See also Orkney Islands.

TANERA BEG

SITUATION: close to Tanera More in the Summer Isles, 3 miles W. of the Ross and Cromarty mainland.
AREA: 267 acres.
POPULATION: uninhabited.
ACCESS: by boat from Achiltibuie.

TANERA MORE

SITUATION: largest of the Summer Isles, 2 miles W. of the Ross and Cromarty mainland.
AREA: 1¼ sq. miles.
POPULATION: uninhabited.
ACCESS: by boat from Achiltibuie.

At the turn of the century seventy people lived on Tanera More, some engaged in crofting and others in the herring-curing industry, a curing station having been set up here. The last islander left in 1946.

Tanera More is a fertile island, with excellent grazing land and much interesting plant life. Dr Fraser Darling, the celebrated naturalist and author of the *Natural History of the Highlands and Islands*, lived on this island for many years and wrote extensively about the Summer Isles (q.v.).

Book : *Island Years*, by Dr Fraser Darling.

TARANSAY

SITUATION: 2 miles W. of Harris in the Outer Hebrides.
AREA: 4 miles by 2 miles.
POPULATION: 5.
ACCESS: by boat from Harris by special arrangement.

The 'special arrangements' for getting to Taransay are such that one voyages there hopefully rather than positively. Instructions issued locally read: 'Tarbert to Horgabust 12 miles. Try to hire the Revd Macdonald's rowing boat. If he won't lend it, light bonfire on headland and hope that the Campbells will pick you up.'

A few years ago an essay on mildewed paper was found in an abandoned schoolhouse on Taransay. It had been composed by a small boy who lived there in the thirties, and wrote:

> winter is the worst season of the year. Here we feel the winter very lonely and dreary when nothing remains of the beauties of time past. Flowers have faded away and birds migrated to warmer climes. . . . At times we do not receive mails for ten or fourteen days, as no boats will venture across the windswept sound. . . . Sometimes we go searching for driftwood along the shore and very often we get a great deal. . . . In winter we see plenty of geese and sometimes swans on the wing which is a sure sign of snow.

It is an accurate picture of life in this outpost. Gale-force winds sweep Taransay in winter and there is little or no protection. The Sound of Taransay can then become as impassable as the seas around Cape Horn. There are only three houses on the island, at the tiny settlement called Paible on the E. side. The old school-house by the landing stage is now used as a store for lobster pots and fishing gear.

The houses have walls of dry stone because there is no lime on Taransay, and the spaces between the stones are filled with earth to keep the wind out. The walls are up to 6 ft thick, and the roof beams are made of drift-wood, there being no trees on the island.

Lobster fishing is the chief occupation, but some sheep are kept on the island and barley, rye, and cabbages are grown. Bleak though it is in winter, Taransay is attractive in summer with its low hills and two small trout lochs: the only disfigurements are the remains of roofless houses long since abandoned.

TEXA

SITUATION: 1 mile S. of Islay in the Inner Hebrides.
AREA: 1 mile by ¼ mile.
POPULATION: uninhabited.
ACCESS: by boat from Port Ellen, Islay.

This heather- and grass-covered island rises to 164 ft at its E. end, and has a small white beacon on its summit. Rocks, reaching a height of 15 ft, extend 7 cables off the SW. corner of Texa.

THAMNA SGEIR: *see under* Sulasgeir.

TIREE

SITUATION: 2 miles SW. of Coll in the Inner Hebrides and 14 miles W. of Mull.
AREA: 29 sq. miles.
POPULATION: 950 (1966).
ACCESS: by air from Glasgow to Scarinish on the E. coast of Tiree, or by boat from Coll.

The most westerly of all the Inner Hebrides, Tiree is a long, flat island, favoured with a remarkably mild climate, as a result of which the inhabitants, most of whom are Gaelic-speaking, can carry on crofting, farming, and even bulb-growing.

Its original name was Tir Eth, meaning 'the land of corn', a reputation it gained in the 6th century when the early Christian missionaries came to the island. Then its flat fertile terrain supplied corn for Iona and other islands. But today Tiree concentrates more on crofting and rearing cattle, and many of the male inhabitants are fishermen. One of the peculiarities of the Inner Hebrides is that an almost freakish mildness of climate is often found a relatively short distance from a bleak, wind-swept and rain-swept island. Tiree in this respect is much more blessed than nearby Coll, and the long hours of sunshine enjoyed on the island have enabled tulips and daffodils to be grown successfully and exported to the mainland.

Thus the island has in recent years basked in a degree of prosperity compared with many others of the Inner Hebrides, where the tendency has been for the people to emigrate and for the local economy to dwindle. Visitors have been encouraged to come to Tiree, and a regular air service has been established to bring in tourists and holiday-makers. Apart from being a green and attractive island with good beaches, Tiree offers visitors a nine-hole golf course and scope for fishing.

Those who wish to make a more leisurely trip to Tiree can do so by taking a boat from Oban to Mull, and going

Tiree (Scottish Tourist Board)

on from there by boat to Tiree.

At one time Tiree was part of the Lordship of the Isles, but in the 15th century the Macdonalds handed it over to the Macleans of Duart, and finally the Campbells of Lochawe took over the island. There are prehistoric duns and *brochs* on Tiree, built by the original inhabitants as defences against Norse and other invaders. From an archaeologist's point of view the most striking of these is Dun Mor Vaul, to the W. of Vaul Bay. This dun, or miniature fortress, is 35 ft in diameter, and its walls are 13 ft thick: it has the remains of cells set into the walls, and galleries.

The principal township is Scarinish, where the airport is situated, and Hynish on the SE. coast is the other settlement.

Tiree is sometimes referred to as 'the island whose heights are lower than the waves'.

Accommodation: there is a hotel at Scarinish, and a number of the inhabitants take paying guests.

TRESHNISH ISLES

SITUATION: $3\frac{1}{2}$ to $7\frac{1}{2}$ miles W. of Mull in the Inner Hebrides.
AREA: 300 acres.
POPULATION: uninhabited.
ACCESS: by boat from Achloch ($6\frac{1}{2}$ miles), from Iona (9 miles), or from Oban (45 miles).

The main islands in this group are Lunga (the largest), Fladda, Dutchman's Cap, and Ciarn A'Burg. There are also twenty-three tiny isles, and many more miniature rocks lying like stepping stones across the sea. The formation of rocks here is volcanic, and most of the islands have steep cliffs rising perpendicularly from the sea, but with smooth, flat surfaces. These cliffs lie on a lava platform near sea level, and in places this platform juts out into the sea and forms a breeding ground for seals.

Fulmars, kittiwakes, puffins, curlews, shags, and shelduck are to be found on some of the islands, and there are also buzzards to cope with the vast rabbit population. The seals are, however, the main attraction, especially the Atlantic grey seal, which can be studied at close quarters.

Transport to the Treshnish Isles is not always easily obtained, and on many islands a landing is impossible. It is sometimes easier to get a boat to the islands from Iona than from nearby Mull. If you have your own boat, it is best to seek local advice on anchoring and landing. Those anxious to study the seals should take a camera, as it is possible to get excellent close-up pictures of the animals.

For lovers of wild life and complete solitude the Treshnish Isles are to be commended.

See also Ciarn A'Burg, Dutchman's Cap, Fladda, Lunga, Sgeir Am Fheoir.

Books: *Island Years*, by J. Fraser Darling, G. Bell & Sons, 1944; *A Tangle of Islands*, by L. R. Higgins, Robert Hale, 1971.

TRODDAY

SITUATION: 2 miles N. of the northernmost tip of Skye in the Inner Hebrides.
AREA: 90 acres.
POPULATION: uninhabited.
ACCESS: by boat from Skye.

The island is used for sheep grazing.

TRONDRA

SITUATION: $\frac{1}{4}$ mile SW. of Scalloway in the Shetland Isles.
AREA: 1 sq. mile.
POPULATION: 20.
ACCESS: by boat from Scalloway.

Green but somewhat barren, Trondra nevertheless formerly supported a population of 100 (1828). Unlike most of the Shetland Isles (q.v.), it has no fishing trade and no harbour.

TUBHARD

SITUATION: close inshore at the mouth of Loch Shell on the E. coast of Lewis.
AREA: 2 miles by ½ mile.
POPULATION: uninhabited.
ACCESS: by boat from Lewis.

ULVA

SITUATION: 150 yds W. of Torloisk, Mull, in the Inner Hebrides.
AREA: 4½ miles by 2 miles.
POPULATION: 45.
ACCESS: by ferry from Torloisk.

Ulva is the largest of all the islands off the coast of Mull. It is also known as Wolf Island, for which animal *ulva* is the Norse name. The island rises from the sea in terraces of basalt to a height of 1,025 ft on the W. side. It has an indented coastline and a large sea cave which runs inland for ¼ mile.

At one time Ulva had a population of more than 800, but after disastrous potato-crop failures in the middle of the 19th century the number of inhabitants dwindled quickly. Some of the best potatoes in the Hebrides were produced here at one time, and the remains of old cottages testify to the one-time prosperity of the island. The inhabitants were often the victims of cruel evictions by landlords.

For 800 years Ulva was owned by the clan MacQuarrie, whose chief entertained Dr Johnson and Boswell when they visited the Hebrides. One of the MacQuarries became Governor-General of New South Wales. Ulva House, a large modern mansion, is situated close to the old cottage where Johnson and Boswell received hospitality. There is also a church on Ulva, built in 1827.

The island is mostly moorland, and covered with bracken in its valleys. It is a private estate today and permission to land has to be sought from the owner. There is a pier where boats can land, and a road runs along the N. shore and connects Ulva with the neighbouring island of Gometra by a bridge.

UNST

SITUATION: the most northerly of the Shetlands, ½ mile E. of the north tip of Yell and 3 miles N. of Fetlar.

AREA: 47 sq. miles.

POPULATION: 1,148.

ACCESS: by inter-island steamer from Lerwick, Shetland Mainland, or by boat from Yell.

Unst is about the same size as Jersey, but with only a fiftieth of the population of that island. It is 12 miles from N. to S. The two principal townships are at Baltasound, where there is a harbour, and Haroldswick on the NE., which has the most northerly post office in the British Isles.

The inhabitants divide their time between fishing and crofting, and one of the island's great advantages is that it has an excellent harbour at Baltasound, another at Uyeasound, and bays at Haroldswick and Norwick which provide shelter for shipping. The highest point of Unst is Saxa Vord on the E. (935 ft), while Hermaness Hill on the W. rises to 657 ft. To the N. of Unst lie the Muckle Flugga Rocks (q.v.), on one of which stands a lighthouse.

Though bleak and less picturesque than some of the other Shetland Isles (q.v.), Unst has reasonably fertile land, and a series of lochs running from N. to S., one of which, Loch Cliff in the N., is the longest if not the

Unst, in the Shetlands (J. Allan Cash)

largest in all the Shetlands. There is abundant fishing, both in the sea and the lochs. Sheep, cattle, and ponies are raised here, and electric power and a piped water supply add to the amenities of the island. The presence of the Royal Air Force at Haroldswick and Saxa Vord has contributed in some measure to improvements on Unst.

Weaving is a secondary industry here, and its quality is such that it is probably the best in all the Shetlands. Hermaness is a nature reserve where the bonxie, or great skua, has found sanctuary. The ruin of Muness Castle in the SE. of Unst is a reminder of the Stewart earls who built it. It is an oblong building with circular towers, and suggests a much greater sense of design than is to be found in most northern buildings of its type. An inscription on the SW. wall of Muness Castle states: 'List ye to knaw this building quha began Laurence the Bruce he was that worthy man quha ernestly his airis and ofspring prayis to help and not to hurt this vark aluyais. The yeir of God 1598.'

The castle was burnt in the 17th century by someone known as Hakki of Dukkeram, but the details of this are obscure. Local legend has it that a young member of the Bruce family abducted a girl named Helga and imprisoned her in the castle. She escaped by a rope from an upper room and the castle was burnt as an act of revenge by her rescuers.

There are said to be ruins of seven churches or chapels on the island, and J. R. Tudor, the authority on Shetland, says that at one time there were twenty-two such buildings. One church was dedicated to St Sunniva, a 10th-century saint who is said to have sailed from Ireland to Norway in a ship filled with virgins. One of the island ships was named after St Sunniva.

Cairns and *brochs* can be found in various parts of Unst; by the Loch of Bordastubble are the Standing Stones, and on the moors known as Crussa Field are the remains of a stone circle. Geologically, Unst is interesting because of its extensive deposits of talc and chromite. Under the ridge of Valla Field is a coarse-grained gneiss, and at the E. of the ridge are masses of gabbro and serpentine. The chromate of iron was formerly quarried at various sites in the E. of Unst.

The population here is still declining, but the inhabitants of Unst have a reputation for hospitality even if accommodation for visitors is scant. Balta Sound, once the scene of great fishing activity, is today a silent reminder of the days when it was a centre of the herring-curing industry.

In the N. of Unst some superb cliff scenery is spoiled only by the radar station which spreads its mast and antennae to the skies.

Accommodation: very limited, but some cottagers take visitors.

Books: *The Orkneys and Shetlands*, by J. R. Tudor; *The Shetland Isles*, by A. T. Cluness.

URIE LINGEY

SITUATION: between Unst and Fetlar in the Shetlands, $1\frac{3}{4}$ miles N. of Fetlar.
AREA: 55 acres.
POPULATION: uninhabited.
ACCESS: by boat from Fetlar or Unst.

UYEA

SITUATION: $\frac{1}{2}$ mile off the SE. tip of Unst in the Shetlands.
AREA: 598 acres.
POPULATION: uninhabited.
ACCESS: by boat from Uyeasound, Unst.

Uyea was inhabited from ancient times until only a few years ago, and is remarkably fertile. It gives protection to Uyeasound in Unst.

The island is low-lying and was at one time the site of a Norse settlement. There are the remains of a pre-Reformation chapel here.

See also Shetland Isles.

UYNAREY

SITUATION: in Yell Sound between Yell and Shetland Mainland.
AREA: 71 acres.
POPULATION: uninhabited.
ACCESS: by boat from Shetland Mainland.

VAILA

SITUATION: 1 mile W. of Shetland Mainland.
AREA: $1\frac{1}{8}$ sq. miles.
POPULATION: 9 (1966).
ACCESS: by boat from Walls, Shetland Mainland.

This is a small but hilly island, enhanced by some diminutive but delightfully green valleys. It also possesses a handsome stone mansion, with a baronial hall and library.

It was on Vaila in 1837 that Arthur Anderson, one of Shetland's most remarkable sons, established the Shetland Fishery Company, thereby breaking the unjust feudal system that had for so many years shackled local fishermen, forcing them to work entirely for the laird's tenant-in-chief. Previously six-oared boats had been purchased from Norway for the local fishermen. Anderson changed all that, and enabled the Shetlanders to build their own craft.

Vaila provides shelter for the Mainland harbour of Walls. Its most prominent landmark is a watch-tower.

See also Shetland Isles.

VATERSAY

SITUATION: $\frac{1}{2}$ mile S. of the isle of Barra in the Outer Hebrides.
AREA: 3 miles by 1 mile.
POPULATION: 104 (1966).
ACCESS: by boat from Barra.

This starfish-shaped island, green, rock-encrusted, with some of the whitest sands in the British Isles, despite its remoteness possesses some thirty-six houses, almost all of them single-storey. It is beautiful but bleak. You can fly to within a few miles of it, but cannot reach it unless the boat can safely cross the wind-whipped sound which separates it from Barra.

Vatersay was colonised by people from nearby islands, notably Mingulay (q.v.), only sixty-five years ago. The original occupiers, known as the 'Vatersay Raiders', built wood cabins and refused to move, despite the threats of the owner, Lady Cathcart Gordon, who somewhat feebly argued that there was insufficient water to support inhabitants. She herself had not even visited the island for half a century. The 'Raiders' engaged a solicitor, fought their case, and were sentenced to six months' imprisonment for illegal occupation of the island; but the pleas of their solicitor resulted in their being let out after six weeks. The 'Vatersay Raiders' became heroes and photographs of them with their solicitor can be seen in many island cottages today.

The pioneer spirit of the 'Raiders' has persisted even though their numbers have been halved in the past decade. Today there is a post office with a telephone box on the island, and even a school attended by less than a dozen children. At the age of eleven they are sent to school on Barra, where they board, returning home each weekend. The people all speak Gaelic and revere the memory of Duncan Campbell, the leader of the 'Vatersay Raiders'. Most of the inhabitants are Catholics.

One road exists on the island, connecting the villagers with their school and the Church of Our Lady of Perpetual Succour at Uidh, where the post-boat lands from Barra. When an islander dies, the body is laid out by the women, who dress it in an embroidered shroud. A wake is held and the Rosary is said at midnight.

The only crops are potatoes, carrots, and oats. Seaweed is used as a fertiliser. A few of the men fish, and some breed and sell black-faced sheep and cattle.

Bleak as Vatersay is, in spring and early summer it is a carnival of flowers. There are primroses in profusion, nestling in every hollow and around every rock, violets, daisies, kingcups, and celandines. Even more remarkable for a treeless isle is the sound of the cuckoo in May. Solans, puffins, and starlings are among the birds to be found there.

Not far from Vatersay Bay a granite monument commemorates the disaster of the *Annie Jane*, which sank in the bay on 28 September 1853, carrying emigrants from Liverpool to Quebec. The monument records that 'three-fourths of the crew and passengers, numbering about 350, men, women and children, were drowned and their bodies interred here.'

Books: *The Enchanted Isles*, by Alasdair Alpin MacGregor, Michael Joseph, 1967; *Summer Days among the Western Isles*, by the same author, Thomas Nelson, 1929.

VEMENTRY

SITUATION: in St Magnus Bay, off Shetland Mainland, ½ mile from shore.
AREA: 1½ sq. miles.
POPULATION: uninhabited.
ACCESS: by boat from Hillside.

This is easily the largest of the uninhabited islands of the Shetlands (q.v.), and if you want to study an all-embracing Shetland landscape, complete with bays, voes, lochs, headlands, and holms, here you can do it undisturbed by human life. It is a splendidly varied landscape with much bird life, including curlew and golden plover, terns, gulls, whimbrel, and wheatear.

Vementry guards the approaches to the narrow sound of Swarbacks Minn, and there have been occasions when its position has been of strategic importance. Consequently in World War I six-inch guns were mounted on the island. They can still be seen today.

The highest point on Vementry is Muckle Ward, a hill which rises to 298 ft. There are cairns on the slopes of the hill which testify to the ancient occupation of this island. Elsewhere on Vementry are ruined *brochs*. It was certainly inhabited more than two thousand years ago, and people were still living there in 1828.

WEST BURRA: *see* Burra.

WEST CARN: *see under* Carn Isles.

WEST LINGA

SITUATION: 2 miles from Shetland Mainland and 1 mile from Whalsay.
AREA: 315 acres.
POPULATION: uninhabited.
ACCESS: by boat from Shetland Mainland.

WESTRAY

SITUATION: 7 miles N. of Rousay and 11 miles W. of Sanday in the Orkney Islands, and 20 miles from Kirkwall.
AREA: 18 sq. miles.
POPULATION: 1,507.
ACCESS: by boat from Orkney Mainland or from Rousay.

Some of the physical features of the inhabitants of Westray are said to be traceable to the crew of a Spanish Armada vessel once sunk off the island. Whatever the truth of this, Westray has characteristics not to be found elsewhere in the Orkneys (q.v.), including a pro-Jacobite tradition that lasted until comparatively recent times. A relic of this is the 'Gentleman's Cave' on the island to which Jacobites retired to drink the health of the 'King over the water' and to escape from the attention of Government informers.

Westray is a mixture of high land and low land, in contrast with most other Orkney islands which are either high or low. Its western coast has a magnificent line of cliffs from the top of which one can obtain some of the finest views in the islands. A distinctive landmark is the ruin of Noltland Castle, built by the Bishop of Orkney, Thomas Tulloch, in 1422. His initials and the figure of a kneeling bishop are carved on the capital of the pillar supporting what must once have been a magnificent winding staircase. The castle was a bishop's residence until 1560, when it passed to the laird, Gilbert Balfour. In the 1745 Rebellion his ancestors supported the Young Pretender and Government troops burnt the castle.

Westwards about 2½ miles from the castle is Noup Head, with its overhanging cliffs swarming with all manner of sea birds, and well worth a visit. 'Gentleman's Cave' is on the W. coast, and it was here that one of the Balfours and his friends hid for a whole winter after their flight from the rout of the Young Pretender's forces at Culloden.

There are piers at Pierowall, the chief settlement on the NE. of Westray, which is also the centre for lobster fishing. In 1968 this was turned to good effect by the creation of a crab- and lobster-processing factory on the island.

Accommodation: limited, and confined mainly to Pierowall.

Book: *Report of the Survey of the Islands of Westray and Papa Westray*, by J. R. Coull, Scottish Development Department.

WHALSAY

SITUATION: 3 miles E. of Shetland Mainland.
AREA: 8 sq. miles.
POPULATION: 764.
ACCESS: by boat from Laxo or Skellister, Shetland Mainland, or by inter-island steamer from Lerwick.

Some of the hardiest and most highly-skilled fishermen in the British Isles live in Whalsay, which is almost entirely dependent upon fishing for a living. The coast around here and the many reefs and shoals in the vicinity are rich fishing grounds, providing an abundance of haddock, herring, and white fish generally.

The island rises to a height of 400 ft at the Ward of Clett, and contains no fewer than eight lochs. There is no crofting or cattle-rearing here, but most of the cottagers have gardens which yield some vital vegetables and fruit. Fresh water and electricity are supplied. The population is concentrated mainly on the W. coast from Symbister to Challister, but there is also a hamlet at Ibister on the E. An enlarged and improved harbour has been developed at Symbister.

Whalsay was inhabited from ancient times, and relics of early occupation have been unearthed there from time to time. Perhaps the most distinctive landmark, however, is the grandiose early-19th-century architectual extravagance – some would call it a monstrosity – known as Symbister House. It was built in a fit of pique by a Shetland laird who decided that, rather than leave his money to relatives he apparently detested, he would spend it on a house that would be a monument to his memory.

There is excellent pasture land on Whalsay, but nobody bothers to use it for the simple reason that fishing provides a far more profitable living.

Off this coast in 1664 a Dutch East Indiaman, the *Carmelan*, was wrecked and broken to pieces on the rocks, and its cargo of three million guilders lost. Some of these coins have been recovered, but so far the bulk of the treasure remains hidden on the sea-bed.

Accommodation: provided at Symbister on a modest scale.

WHITE ISLAND, THE: *see* Eilean Ban.

WIAY (Benbecula)

SITUATION: 1 mile off the SE. tip of Benbecula, Outer Hebrides.
AREA: 2 miles by 1–2 miles.
POPULATION: uninhabited.
ACCESS: by boat from Benbecula.

This island should not be confused with another Scottish island of the same name off the coast of Skye. It is surrounded by some seven islets S. of Benbecula, situated in the strait known as Bagh Nam Faoleann between Benbecula and South Uist. Most of them are only a few acres in size.

WIAY (Skye)

SITUATION: in Loch Bracadale, 2 miles W. of the coast of Skye, Inner Hebrides.
AREA: 1½ miles by 1 mile.
POPULATION: uninhabited.
ACCESS: by boat from Skye.

The island is used for grazing. There is excellent fishing in the vicinity.

WOLF ISLAND: *see* Ulva.

WYRE

SITUATION: 1 mile SE. of Rousay in the Orkneys.
AREA: $2\frac{1}{2}$ miles by $\frac{3}{4}$ mile.
POPULATION: 36.
ACCESS: by boat from Rousay or Orkney Mainland.

Excavated remains of a castle built by Kolbein Hruga in the days of the Norse occupation are to be seen on the tiny Wyre Island. It is said that the invaders deliberately chose small islands for such fortifications because they could not be taken by surprise and were easier to guard.

Wyre takes its name from the Norse *vigr*, meaning 'arrowhead', which it resembles in shape. Kolbein Hruga was a warlike chieftain of the 12th century, described in the *Orkneyingers' Sagas* as 'the most haughty of men'. The *Sagas* also refer to the castle on Wyre as 'a safe stronghold' and recall that the murderers of a Norse earl took refuge in it. The castle has a square keep surrounded by a ditch, and traces can be seen of an upper floor, a ground-level chamber, and a water tank. The stone-work is of a comparatively high standard.

Also worth seeing are the remains of a Norse farm, Bu of Wyre, and a roofless church dating from the 12th century. A school exists on the island.

See also Orkney Islands.

YELL

SITUATION: 2½ miles NE. of Shetland Mainland and, at its most northerly point, ½ mile W. of Unst.
AREA: 83 sq. miles.
POPULATION: 1,155 (1966).
ACCESS: by inter-island steamer from Lerwick, or by boat from Mossbank, Shetland Mainland.

Second largest of the Shetland Isles (q.v.), Yell is 17 miles from N. to S. and 7 miles wide at its widest part. Though large, it is not the most interesting or rewarding of the islands, and considerable expanses of it are little more than neglected peat moorland. Its main centre of population is Mid Yell, where there is an admirable natural harbour, sheltered in part at its approaches by the island of Hascosay (q.v.).

More than any other of the Shetlands, Yell has suffered from depopulation and depression. Once it had a population of 2,700, but today it has an appalling unemployment problem, and young people are continually leaving the island. Perhaps one of its greatest problems is that it is an island that lacks leadership: there is no laird or proprietor to give it a plan.

One reaches Yell by way of Ulsta in the S., which is the terminal for the Yell Sound ferry. From Ulsta one road runs N. along the W. coast through some of the best scenery in the islands, and another runs E. through Burravoe and then N. via Gossabrough and Otterswick to Mid Yell.

The voes of Yell resemble the fjords of Norway, and are one of the outstanding attractions of the island. But Yell is still one of the most neglected of the islands, despite its size. Agriculture remains on a crofting level, fishing has been neglected, and little has been done about the abundance of peat on Yell. If ever an island needed a subsidy, if only a temporary subsidy, it is Yell. A report by the Shetland Development Officer in 1962 stated that

'Yell's economy is at present so fragmented and indeterminate that radical improvement will require agreement among the development agencies as to the relative significance of their various fields of action. . . . Today it finds itself in an alien climate.'

Yet the Faeroe Islands, far to the N. of the Shetlands, seem to have solved problems that should equally be solved by Yell. There is a confidence among the Faeroese that is lacking in the inhabitants of Yell: a report on the former stated that they 'possess unlimited confidence in their future. The young folk stay on and the man of means is not afraid to risk his capital in local development.'

The interior of Yell, described by Eric Linklater as 'dull and dark and one large peat-bog', could be developed to a far greater extent. Edward Charlton, a Roman Catholic visitor to Yell, wrote in 1832 that he found the island 'neglected by the Presbyterian ministers', and that the 'poor Shetlanders have taken refuge in the arms of the Wesleyan missionaries and the Methodists now outnumber in these remote islands the followers of the Church of Scotland.'

The coastal part of Yell is much more attractive. But perhaps the most pleasing part of the island is that around Lumbister, where the land is greener and more fertile.

Books: *The General Grievances and Oppressions of the Isles of Orkney and Shetland*, by J. Mackenzie; *The Shetland Isles*, by A. T. Cluness.